Fractals and Dynamic Systems in Geoscience

Edited by
Tom G. Blenkinsop
Jörn H. Kruhl
Miriam Kupková

2000

Birkhäuser Verlag
Basel · Boston · Berlin

Reprint from Pure and Applied Geophysics
(PAGEOPH), Volume 157 (2000), No. 4

Editors:

Tom G. Blenkinsop
Department of Geology
University of Zimbabwe
Harare
Zimbabwe
e-mail: tgb@icon.co.zw

Jörn H. Kruhl
Institut für Angewandte Geologie und
Mineralogie
TU München
München
Deutschland
e-mail: joern.kruhl@geo.tum.de

Miriam Kupková
Institute of Materials Research
Slovak Academy of Sciences
Košice
Slovakia
e-mail: kupkova@imrnov.saske.sk

A CIP catalogue record for this book is available from the Library of Congress, Washington D.C., USA

Deutsche Bibliothek Cataloging-in-Publication Data

Fractals and dynamic systems in geoscience / ed. by Tom G. Blenkisop ... - Basel ; Boston ; Berlin : Birkhäuser, 2000
   (Pageoph topical volumes)
   ISBN 3-7643-6309-6

This work is subject to copyright. All rights are reserved, whether the whole or part of the material is concerned, specifically the rights of translation, reprinting, re-use of illustrations, recitation, broadcasting, reproduction on microfilms or in other ways, and storage in data banks. For any kind of use, permission of the copyright owner must be obtained.

© 2000 Birkhäuser Verlag, P.O.Box 133, CH-4010 Basel, Switzerland
Printed on acid-free paper produced from chlorine-free pulp. TCF ∞
Printed in Germany

9 8 7 6 5 4 3 2 1

Contents

485 Preface
T. G. Blenkinsop, J. H. Kruhl and M. Kupková

487 Appolonian Packing and Fractal Shape of Grains Improving Geomechanical Properties in Engineering Geology
C. A. Hecht

505 Fractal Characterization of Particle Size Distributions in Chromitites from the Great Dyke, Zimbabwe
T. G. Blenkinsop and T. R. C. Fernandes

523 Dynamics and Scaling Characteristics of Shear Crack Propagation
V. V. Silberschmidt

539 Fractal Approach of Structuring by Fragmentation
C. Suteanu, D. Zugravescu and F. Munteanu

559 Micromorphic Continuum and Fractal Fracturing in the Lithosphere
H. Nagahama and R. Teisseyre

575 Spatial Distribution of Aftershocks and the Fractal Structure of Active Fault Systems
K. Nanjo and H. Nagahama

589 Scale-invariance Properties in Seismicity of Southern Apennine Chain (Italy)
V. Lapenna, M. Macchiato, S. Piscitelli and L. Telesca

603 Variation of Permeability with Porosity in Sandstone Diagenesis Interpreted with a Fractal Pore Space Model
H. Pape, C. Clauser and J. Iffland

621 Slow Two-phase Flow in Single Fractures: Fragmentation, Migration, and Fractal Patterns Simulated Using Invasion Percolation Models
G. Wagner, H. Amundsen, U. Oxaal, P. Meakin, J. Feder and T. Jøssang

637 Dynamic Model of the Infiltration Metasomatic Zonation
V. L. Rusinov and V. V. Zhukov

653 Wavelet Analysis of Nonstationary and Chaotic Time Series with an Application to the Climate Change Problem
D. M. Sonechkin and N. M. Datsenko

Preface

Fractal geometry and nonlinear systems theory are increasingly employed in various branches of science. The widening use of these concepts and techniques can break down barriers between different and hitherto relatively isolated fields, and can help to create new interdisciplinary subjects. The geosciences are particularly suitable for the application of fractal geometry and nonlinear systems, because a vast range of examples of fractal geometry and nonlinear processes occur within the subject.

Research in fractal geometry may be presented in many different specialised meetings devoted to other branches of science. One of the aims of the series of international symposia "Fractals and Dynamic Systems in Geosciences" is to bring together people working in different fields of geoscience who deal with the terms and procedures of fractal geometry and nonlinear systems, and to serve as an umbrella under which to discuss different topics from a common viewpoint. Experience has shown that this can be quite fruitful. An interesting beneficial corollary of this conjunction is the generation of new interest in different subdisciplines within the geosciences.

There is still a need to develop more useful tools to analyse the complex structures of fractal geometry and the processes of nonlinear systems, such as self-organisation, which occur widely in the geosciences and are important for understanding natural processes. There is also a need to introduce these concepts and tools to a broader audience in the earth sciences (where there has also been resistance to some of these ideas). The editors hope that this special volume will both advance and disseminate the new scientific approach.

After the first two successful meetings in 1993 and 1995 at Gelnhausen, Germany, the third symposium was held on the same topic at Stara Lesna, Slovakia in June 1997, at which 41 contributions were presented. This thematic issue of Pure and Applied Geophysics presents some of the contributions from the third symposium and some papers called for by the editors from participants of previous symposia. We would like to thank all of the participants for the success of the symposium; the authors for their papers, and the referees for their important contribution. Sponsorship for the conference is gratefully acknowledged from the EAGE-PACE foundation, the Stifterverband für die Deutsche Wissenschaft and Institutes of Slovak Academy of Sciences situated in Košice, Slovakia, namely

Institute of Geotechnics, Institute of Materials Research and Institute of Experimental Physics.

It is particularly our pleasure to thank R. Dmowska, editor-in-chief of topical issues of Pure and Applied Geophysics, for her useful recommendations and patience during the preparation of this volume.

> Prof. Tom G. Blenkinsop
> Department of Geology
> University of Zimbabwe
> P.O. Box MP 167
> Mount Pleasant
> Harare
> Zimbabwe
> E-mail: tgb@icon.co.zw

> Prof. Dr. Jörn H. Kruhl
> Institut für Angewandte Geologie und Mineralogie
> TU München
> Arcisstr. 21
> D-80290 München
> Germany
> E-mail: joern.kruhl@geo.tum.de

> Dr. Miriam Kupková
> Institute of Materials Research
> Slovak Academy of Sciences
> Watsonova 47
> 04353 Košice
> Slovakia
> E-mail: kupkova@imrnov.saske.sk

To access this journal online:
http://www.birkhauser.ch

Appolonian Packing and Fractal Shape of Grains Improving Geomechanical Properties in Engineering Geology

CHRISTIAN A. HECHT[1]

Abstract—Fractal packing and highly irregular shaped particles increase the mechanical properties of rocks and building materials. This suggests that fractal methods are good tools for modeling particle mixes with efficient properties like maximum strength and maximum surface area or minimum porosity and minimum permeability. However gradings and packings are calculated by "Euclidean" disk models and sphere models. Surprisingly even the simplest models are far more complex than they appear. The fractal "Appolonian packing model" is proposed as the most universal two-dimensional packing model. However the inhomogeneity of gradings and the irregularity of natural grain shapes and surfaces are not reflected by these models. Consequently calculations are often far from empirical observations and experimental results. A thorough quantification of packings and gradings is important for many reasons and still a matter of intense investigation and controversial discussion. This study concentrates on fractal models for densely packed non-cohesive rocks, crushed mineral assemblages, concrete and asphalt mixtures. A summary of fractal grain size distributions with linear cumulative curves on log-log plots is presented for these mixtures. It is shown that fractal two-dimensional and three-dimensional models for dense packings reflect different physical processes of material mixing or geological deposition. The results from shear-box experiments on materials with distinct grain size distributions show a remarkable increase of the mechanical strength from non-fractal to fractal mixtures. It is suggested that fractal techniques need more systematical application and correlation with results from material testing experiments in engineering geology. The purpose of future work should lead towards the computability of dense packings of angular particles in three dimensions.

Key words: Appolonian, packing, fractal, engineering, geology.

Introduction

The demands upon the qualities of building materials are increasing rapidly as a result of new building technologies and economic competition. The material properties reach to the limits of the potential of the single natural component or the ideal mixture. Various attempts are made to improve the strength of asphalt mixtures in response to increasing traffic, and to enhance the impermeability and density of high quality concrete. In addition to the strength of binding agents and

---

[1] Martin-Luther-Universität, Halle-Wittenberg, Institut für Geologische Wissenschaften und Geiseltalmuseum, Domstraße 5, 06108 Halle-Saale, Germany, e-mail: hecht@geologie.uni-halle.de

the stability of single particles, the packing density becomes a powerful element of the bulk material strength.

Models for dense packings always followed the hierarchical concept where spaces between the largest first-order particles are filled by second-order particles of exactly the diameter to fit into the pores, followed by third-order particles and so on. It was recognized early that a high density of concrete was achieved by continuous mixtures that accord to the "Fuller-formula" (FULLER and THOMPSON, 1907). Empirical work on building materials like asphalt led to more discontinuous compositions with a high percentage of the coarse grain fraction (ZTV-ASPHALT-StB 94). Later discontinuous mixtures for very dense packings, so-called "gap gradings," were developed and calculated on the basis of circular disk models with disks representing one grain size having equal diameters. It was shown that the calculation of particle diameters and quantities for gap gradings became more realistic if a three-dimensional sphere model was used (KESSLER, 1994). The complex geometry of pore spaces in the latter model was found to make gap gradings indeterminable. As fractals became more popular, the self-inverse "Appolonian disk model" was proposed for dense concrete (KAYE, 1989). Because natural particles are angular and have variable sizes, and particle mixes are not homogeneous, these calculations are still a simple approach to the packing problem.

It is the aim of this paper to identify fractal packings in engineering geology and to elaborate the general aspects of their formation and their potential geomechanical behavior. The study comprises non-cohesive materials that are composed of natural mineral grains from 1 to n orders of sizes. Both technical mixes that cover the scale of particle diameters from 0 mm–100 mm, and natural rocks that may contain large blocks up to 100,000 mm in diameter and more, are taken into consideration.

Although concrete and asphalt mixes are included, the main subjects in view are grain geometries and packings rather than properties of cements and other binding agents. The main aspects of fragmentation, mixing and sedimentation associated with the concept of fractal packing are discussed. The results provide good evidence for the similarity of physical processes that generate dense packings in nature and in the rock industry. It is shown why disk models and sphere models are too complex to allow absolute calculations of densest gradings. An attempt is made to show that Appolonian packing is a uniform two-dimensional disk model limited to certain powerful packings. The experimental results show how shear strength values expressed by the friction angle, φ, reflect a special type of packing and possibly correlate with the fractal dimension D for those examples with fractal grain size distributions. It is suggested that grain shape and surface conditions affect the mechanical material properties at different scales and must be introduced into future packing models.

Observations

Geological Properties of Clastic Soft Rocks

Clastic soft rocks have various depositional histories. Consequently they show an exceptional compositional variety. Modern methods to analyze the composition of rocks cover a large range of scales. High resolution microscopes open more and more two-dimensional insights into rock components. New remote sensing methods allow new views on geological structures on the largest scale. Probably the most prominent result of rock composition analysis in engineering geology is the particle size distribution, which can easily be determined by sieving or gravitational settling methods on a laboratory scale. Methods are becoming more complicated on the large scales above 1000 mm particle diameter and on the small scales below 0.001 mm where there is hardly any information on particle size distributions. If many particle grades are present, techniques with different resolution scales must be combined, which causes calculation problems. Most of the knowledge has been accumulated on the scale of standard laboratory methods and rock engineering. The log-linear cumulative grain size distribution curve is the graphical standard method to visualize the composition of soft rocks. Instead of a linear scaling of the y-axis, the Rosin Rammler Weilburg probability scale (ROSIN *et al.*, 1933), or the log-normal probability scale or a log-scale can be used for statistical analyses of particle distributions. It was shown that the histogram is still very important for the determination of clustering particle domains in solid fragmentation (SUTEANU *et al.*, 1993). Many natural soft rocks show continuous cumulative curves, divided into homogeneous rocks that are composed of a few grain sizes and inhomogeneous rocks that contain many sizes of grains (Fig. 1). Of course fractal patterns most likely occur among the inhomogeneous rocks where many scales are present. For consolidated rocks, grain size distributions can be calculated from particle counting in thin sections (PAPE and SCHOPPER, 1987). If a log-log plot is used, the distributions with a power function, like the "Fuller-formula" produce linear cumulative grain size distribution plots (Fig. 1). These rocks are known to have the lowest possible porosities among continuous distributions. Later it will be shown that disk models and sphere models lead to discontinuous "fractal" distributions that are used in concrete production to further enhance the packing density.

The initial fragmentation of rocks is driven by different natural forces. The most important fragmentation forces for the scales investigated here are mechanical rock weathering and explosive volcanic action. They produce the majority of the initial particle mixes on the earth's surface, many of which apparently are self-similar fractals (TURCOTTE, 1997) with number-size distributions that follow the power-law

$$N = C/r^D \tag{1}$$

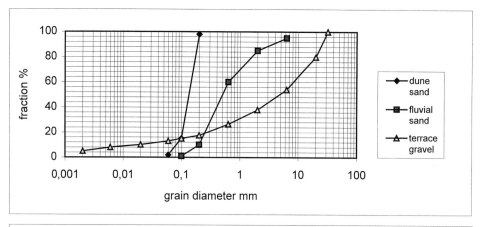

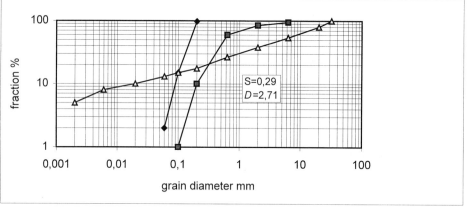

Figure 1
Log-linear and log-log cumulative distribution curves of natural soft rocks with different grades of homogeneity. The linear log-log plot shows the power distribution of the inhomogeneous terrace gravel (log-linear distributions after PRINZ, 1991). S = slope, D = fractal dimension.

with N being the number of particles with a linear size r, C the constant of proportionality and D the fractal dimension.

Each spherical particle has the volume $(4/3)\pi r^3$. Consequently the volume V_r of N particles with a radius r is

$$V_r = N(4/3)\pi r^3 = C(4/3)\pi r^{3-D}. \qquad (2)$$

Normalizing V_r by the total Volume V yields the relative matrix volume $\Phi_{mtx,r}$ corresponding to grain radius r.

$$\Phi_{mtx,r} = V_r/V = C(4/3)\pi r^{3-D}/V = C_2 r^{3-D} \qquad (3)$$

with $C_2 = C(4/3)\pi/V$.

Equation (3) defines the frequency distribution of grain sizes weighted by volume and can be written as

$$\log \Phi_r = \log C_2 + (3 - D)\log r. \qquad (4)$$

Integration of eq. (3) with respect to log r yields the cumulative distribution function $G(r)$

$$G(r) = \int \Phi_r \, d\log r = C_2 \int r^{(3-D)} \, d\log r$$

$$= C_2 (3 - D)^{-1} r^{3-D} + \text{const.} \qquad (5)$$

As $G(0) = 0$, const $= 0$. From this follows

$$G(r) = C_3 r^{(3-D)} \qquad (6)$$

and

$$\log G(r) = \log C_3 + (3 - D)\log r, \qquad (7)$$

with $C_3 = C_2(3 - D)^{-1}$ and $0 < G(r) < 1$.

From this follows

$$r_{\max} = C_3^{-1/(3-D)}. \qquad (8)$$

Equations (4) and (7) show, that in a log-log plot, the frequency distribution and the cumulative distribution of grain sizes weighted by volume both are represented by a straight line with the slope (3-D). The offset on the axis of log $G(r)$ is—ln (3-D).

$G(r)$ corresponds to fraction % in Figures 1, 2, 4, 5 and 6.

The fractal dimension D of fractured rocks has values between 1.9 and 2.9 (TURCOTTE, 1997). The experimental crushing results of this study shown in Figure 2, give fractal dimensions $D = 2.43$ (roll crusher) and $D = 2.32$ (jaw crusher) for a Permian rhyolite. PAPE and SCHOPPER (1987) presented a method for the stereological determination of grain size distributions on the basis of a Sierpinsky sieve with cubic cells. They applied the model to different sandstones and found a large linear region in the log N_a vs log r plot for greywackes with a wide spectrum of grain sizes. SUTEANU et al. (1993) pointed to a clustering of dominant sizes in fragmented "self-similar" rocks that is visible in a histogram but probably hidden under a number-size log-log plot. Although there are exceptions, the initial rock fragments are more or less angular shaped with a rugged surface. Within a certain scale range they are statistically self-similar and have highly irregular shapes and surfaces which makes them good candidates for a determination by fractal methods e.g., the Richardson step divider method (RICHARDSON, 1961). Fragmentation obviously produces fractal patterns irrespective of the type of destructive energy and the mineral composition of the fragmented rockmass.

Rock fragments are transported by gravitational forces or currents either in water or in the air. During transport they can undergo selection and chemical or

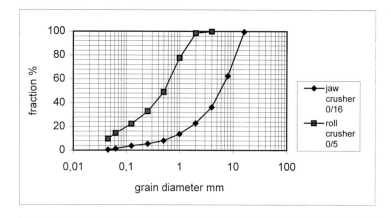

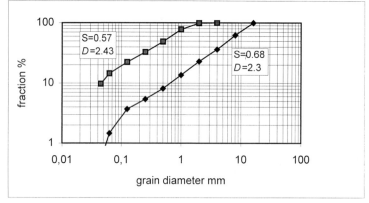

Figure 2
Log-linear and log-log cumulative distribution curves of a crushed volcanic rock with a high quartz content. The fine material (0/5 mm) was processed from the primary crushed coarse grained sample (0/16 mm). The log-log plot shows the power distribution of both materials. S = slope, D = fractal dimension.

mechanical alteration which changes their fractal dimension to a certain degree. The fractal composition of a fragmented rock mass can be strongly reduced through selective transport, which leads to a reduction of the grain distribution to a few sizes of grains. Mechanical alteration can also decrease the total surface area of grains by rounding the grain body and polishing the grain surface. In contrast, chemical alteration can enhance the surface roughness of a grain and thereby the fractal dimension of its surface. Depending on the type of mass transport, particle assemblages are either destroyed or remain more or less connected within the original mass. In the first case the transport energy is selective and separates the original grain assemblage into clusters or rhythmic layers of equal particle size. In the second case the transport energy is conservative and almost preserves the original composition which, as shown before, is fractal in many cases. For example, in a rock fall or debris fall, a fragmented mass is transported at high speeds or free fall over a short horizontal distance. The resulting deposits, such as breccias, often resemble fractal patterns. Other mass transports, like debris flows, also produce unsorted deposits that contain many grain scales. Turbidity currents carry various grain sizes although the deposits are often sorted in a rhythmic pattern well known as the Bouma sequence (BOUMA, 1962). The coarse bottom layers of such sequences, however, often contain many grain scales and show a Fuller-type of composition (KUENEN, 1964). Among grain flow and fluidal flow deposits, polyphase deposits can be formed where different grain sizes are transported and flushed into each other during individual flood events so that the pore spaces are continuously filled by each depositional phase (WASSON, 1977). During this process the large grains would act like a sieve and allow only those particles to infilter that fit through the narrowest triangular wedges between the large grains. Braided stream and fan deposits for example can be formed by high energy multidirectional and periodical flow. Sediments of this type (e.g., fluvial conglomerates and fanglomerates) have fractal packings that are composed of polyshaped "Euclidian and fractal" particles (Fig. 3). Coastal

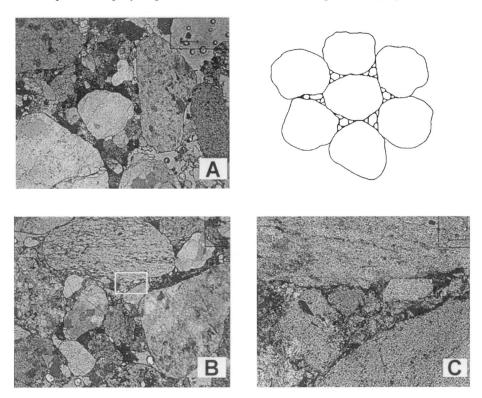

Figure 3

Thin section through a fine conglomerate showing "Appolonian-packing" of rounded and angular clastic grains of mainly quartz. A: ideal natural "Appolonian net" (section length 2.5 mm). B: dense packing with marked inset of Figure 2. C (section length 2.5 mm). C: detail of densely packed interstitial space (section length 0.35 mm). Drawing shows a schematic Appolonian packing of angular grains.

sediments are composed of rounded "Euclidian" particles and reach a low range of different grain sizes. They exhibit high porosities and good permeabilities. Wind deposits are even better sorted however, the particle surfaces are rather more speckled with small indentations than polished through transport.

The following table summarizes the main types of deposition and their potential to form fractal grain size distributions and fractal packings. The first column contains the main types of clastic deposition on the basis of transport energy. In the second column, the potentials for fractal grain size distributions and particle irregularities are estimated from the general types of grain size distributions and general particle geometries. Column three contains rock examples with numbers for the fractal dimension D given for those grain size distributions characterized by true linear or mainly linear graphs on number vs size or cumulative distribution log-log plots. Citations are listed for those rock types where fractal dimensions have been published for special examples. Some of the published log-linear distribution data have been transformed to log-log plots in this study. The fractal dimensions derived from this examples are presented in Figures 1, 2 and 4–6 and Table 1 below.

Geomechanical Properties of Natural Soft Rocks

The geomechanical properties of non-cohesive soft rocks depend strongly on grain shape, grain size distribution and packing. Porosity and permeability studies of rocks envisage the pore space geometry which is fractal in many cases (PAPE et al., 1984) rather than the grain framework itself. The latter, however, is more

Table 1

Summary of main types of deposition and their potential to form fractal grain size distributions and fractal packings

Type of deposition	Potential for fractal grain size distribution/particle irregularity	Rock example/fractal dimensions D of grain size distribution	Reference
Rock fall	high	Breccia[1]	
Debris fall	high	Fractured rock[1]	
Rock slide	high	Fractured rock mass[2]	This study Fig. 1
Polyphase flow	high	Terrace gravel 2.7–2.8	
Debris flow	high-medium	Matrix supported gravel	
Turbidity current	medium-low	Greywacke 2.8	PAPE and SCHOPPER (1987)
Grain flow	low	Fluvial sand	This study Fig. 1
Wave action	very low	Coastal sand	
Wind	extremely low	Dune sand	This study Fig. 1

[1]Fractal dimension D characterizes a disturbed fragmented rock mass. [2]Fractal dimension D characterizes an almost nondisturbed fragmented rockmass.

important for the shear strength of non-cohesive soft rocks. This shear strength is composed of friction and dilatancy. Friction depends on the roughness of the grain surfaces and the contact area between grains. The standard roughness analysis in engineering geology is descriptive and scale dependent (BAMFORD, 1978). Grain boundaries and grain shapes can also be determined by estimation of their fractal dimension by the step divider method (KAYE, 1989; LEE et al., 1990). The fractal dimension is a number directly correlated to grain shape or surface roughness, and can be correlated with geomechanical parameters or be introduced into calculations and models. Dilatancy is dependent on the space available to grains that are forced to move over each other during shearing, which is usually calculated by the relative density of a rock. It is well known from empirical results that the shear strength of inhomogeneous soft rocks is higher than that of homogeneous rocks (ORTIGAO, 1995). The amount of dilatancy depends on the type of packing of a rock. Theoretically the highest density can be reached in fractally composed rocks where all pores are occupied by ideally fitting grains (Fig. 3). Therefore the fractal density of the packing is an interesting parameter in addition to the bulk density of a rock.

Table 2 presents the results of shear-box experiments carried out on three calculated laboratory mixtures of a weathered granite. The grains consist mainly of quartz and feldspar which are angular shaped. The grain size distributions range from 0 mm to 8 mm with a content of the fractions smaller than 0.063 mm of less than 5%. The three types of distributions selected for this experiment were: firstly, a homogeneous distribution comparable to the example of a fluvial sand in Figure 1; secondly, a mixture calculated by the Fuller formula; and thirdly, a gap grading comparable to the examples of Figure 6. The shear-box experiments were carried out in three steps with confining pressures of 0.1 MPa, 0.35 MPa and 0.75 MPa, respectively. The experiments were repeated for three sample conditions: dry air, 10 mass % water content, and water saturation. The table gives the friction angles, φ in degrees for the different samples.

As was discussed earlier, the packing density of the grain sizes continuously increases from a homogeneous distribution to a Fuller distribution and finally to a gap grading distribution. The mechanical properties related to the packing density

Table 2

Results of friction angles φ in degrees from shear base experiments carried out on three calculated laboratory mixtures

Sample condition Fractal dimension	Homogeneous distribution not fractal	Fuller distribution $D = 2.4$–2.5	Gap grading $D = 2.5$–2.6
Dry air	28	44	46.6
10 mass % water	33	49.3	56
Water saturation	29.5	47	54

also change systematically. The results show an increase of the shear strength expressed as the friction angle, φ. The difference between the continuous mixture and the Fuller mixture is large, which is expected, but it is still significant between the Fuller mixture and the gap grading mixture. The results from experiments on natural soft rocks show the same trend, although their compositons vary widely The fractal dimension is $D = 2.44$ for the Fuller distributions of sand to gravel-sized materials calculated from the log-log plot in Figure 4. The fractal dimension for the gap grading material is difficult to calculate since the graphs are not linear over the full range of grain sizes. The dimensions calculated for the linear portions of the graphs in Figure 6 give values of $D = 2.5–2.6$, which are higher than those for the Fuller distributions. Future systematic work will be conducted to relate fractal dimension values, D, derived from log-log cumulative distribution curves, to mechanical parameters, the friction angle in this case.

General Properties of Mineral Mixes

Many of the basic observations on natural rocks can also be made on technical mineral mixes. This is not surprising because they are mixtures of natural soft rocks, for example gravel and sand, of crushed hard rocks or of a combination of both. Regardless of their geological origin, particle distributions and particle properties of building materials are selected solely from geotechnical and economic points of view, which is a straightforward approach as opposed to the variety of soft rocks produced by various natural forces.

It is well known from the mining industry that fragmentation of rocks through blasting, crushing or grinding is often related to a power-law which is similar to the observations of rock fragmentation through natural forces (Fig. 2). The techniques of blasting and crushing depend on the type of rock and the intended use of the crushed material. High quality gradings are double crushed and sieved so that a standardized material contains hard grains of regular shape and equal size. Ideal grains are angular but not platy or elongate, which are unsuited for dense packings. The grain shape is simply measured as the ratio between long and short axes. It expresses the elliptical shape of the grain envelope. However, the axial ratio tells one nothing about the ruggedness of the grain or its surface condition. In concrete or asphalt, the surface of mineral particles becomes more important as a reaction area for the binding agents. Crushed grains are generally angular but exhibit different surface conditions. Rounded grains often have smooth surfaces and small total surface areas. Very smooth to glassy surfaces are useless in asphalt because the bitumen runs off the grain surfaces during the mixing process. On the other hand, very large surface areas consume considerable binding agent, which may result in a higher strength, but simultaneously increases the price of the mixture. As the volume of grains is finite, the fractal surface area is infinite in theory if the inspection scale is unlimited. Fractal techniques have proved very useful for particle analysis in many fields, although the scales of resolution are limited in practice (KAYE, 1989). The fractal dimension of the surface can be approximated by a step divider calculation of the fractal dimension D of a one-dimensional contour trace plus unity:

$$D_{contour} = \log N(r)/\log r \quad \text{and} \quad D_{surface} = D_{contour} + 1 \tag{9}$$

where $N(r)$ is the number of steps of the size r, $D_{contour}$ is the fractal dimension of the grain contour and $D_{surface}$ the fractal dimension of the grain surface (TAKAYASU, 1989).

Mixing energies in plants or on site are generally high compared to natural processes. From a technical point of view the production of homogeneous mixtures presents no problem if the component sizes are standardized and the recipes are well calculated. As opposed to natural rocks, building materials are presumed to be transported without much disturbance. Demixing and sorting happen mainly to non-cohesive coarse materials when dynamic energy is added during transport, e.g., by vibration or by a fast destruction of potential energy during loading and piling. Another important aspect of mixing is the workability of fresh concrete, expressed by the flow ability. The latter is hindered by a high grain angularity of the coarse fractions, which produces a high internal friction of the mixture. Concrete mixtures with gap gradings are also known to have a limited workability as a result of high interparticle friction and dilatancy between the coarse grain fraction, which corresponds well to the results of Table 2.

Packing Models

Two fundamental types of cumulative distribution curves have been developed for dense packings. They are either continuous or discontinuous, however both are obviously based on fractal concepts although they are remarkably distinct. Perhaps the most prominent continuous curve is the Fuller distribution which has been used as a standard recipe for many years because it is easily calculable (Fig. 4). In order to further optimize packing densities, the cumulative curves have been modified and shifted towards an unsteady type of distribution, for example in asphalt mixtures for roads with heavy traffic (Fig. 5, top layer asphalt). Unsteady grading curves, so-called "gap gradings," have been developed for very dense concrete and also partly for asphalt mixtures (Fig. 6). The log-log plots of the cumulative curves are linear for the Fuller distributions (Fig. 4) but not for the gradings of the top layer asphalt (Fig. 5) or for the gap gradings (Fig. 6). The slopes of the latter have been calculated for the linear

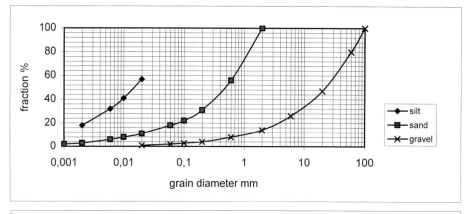

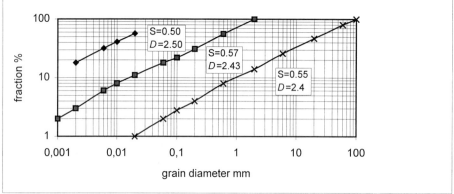

Figure 4
Log-linear and log-log cumulative distribution curves of non-cohesive rocks with Fuller composition (log-linear distribution after STEFFEN, 1986). S = slope, D = fractal dimension.

portions of the graphs. If the point for the gap grain size were to be eliminated from the graph of the log-log plot than an almost linear plot is produced on the right side. The fractal dimension, D, calculated by 3-S (S = slope) is higher for the gap gradings ($D = 2.5-2.6$) than those calculated for the Fuller distributions ($D = 2.43-2.5$).

Gap gradings are calculated on the basis of simple circular disk or sphere models (Fig. 7). KESSLER (1994) showed that the calculations prove very difficult even though "Euclidean" particle geometries are used in the models. The radii calculated from two-dimensional disk models are underestimated because the interstitial spaces in a three-dimensional sphere model are larger than in the former models (Fig. 7,B). Although the basic equations were corrected, KESSLER (1994) pointed out that for a number of reasons the calculation of an ideal gap grading is impossible. One of the reasons was the observation that the tetrahe-

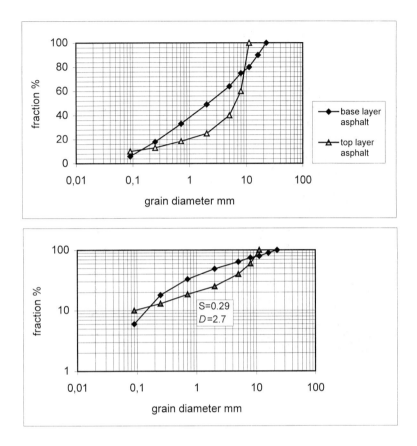

Figure 5
Different asphalt mixtures: Base layer asphalt, texture of large grains separated by smaller grain fractions. Top layer "splittmastix" asphalt, very dense "fractal" texture with contacts between the oarse grains and filling of the interstitial spaces by the fine fraction (log-linear distributions after ZTV-Asphalt-StB 94). S = slope, D = fractal dimension.

dral and octahedral spaces in his model become very complex from the third order of grains downwards. Another model, better known to fractalists, is the "Appolonian net" (Fig. 7,C) which is also a two-dimensional disk model. It was described by MANDELBROT (1991) as a complex self-inverse fractal. He demonstrated that the interparticle triangles of the Appolonian net are not scale-invariant and not self-similar. This observation also reflects the complex wedge geometries produced by neighboring disks which also have various diameters in the Appolonian net.

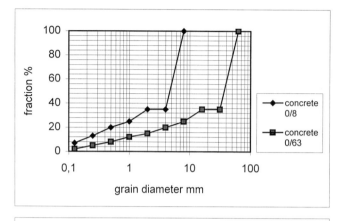
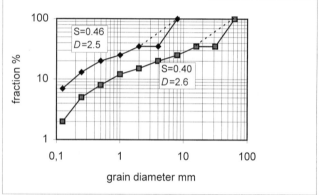

Figure 6
Log-linear and log-log cumulative distribution curves of concrete "gap gradings" showing the dominance of the first-order coarse grain fraction (log-linear distributions after KESSLER, 1994). The cumulative curves reflect the fractal "Appolonian packing" concept of the gradings. S = slope, D = fractal dimension. S and D were calculated from the linear portion of the graph.

Discussion

Rocks are not fractal "sensu stricto" from a mathematical point of view because they do not cover an infinitely long range of scales, although the methods of observation are tending towards both ends of the scale. Nevertheless, natural rocks can be analyzed by fractal methods if they are rugged and complex enough (MANDELBROT, 1991). Engineering geologists need to predict the geomechanical behavior of particles or rock masses. They introduce well-defined elements like buildings or technologies into the systems and consequently they must make geology computable. The majority of geological classifications is still descriptive and the applied models use "Euclidean" geometries. As natural shapes are irregular, their geometries are reduced by the latter models and the geotechnical numbers become arbitrary. Since it was found that fractal techniques give numbers to

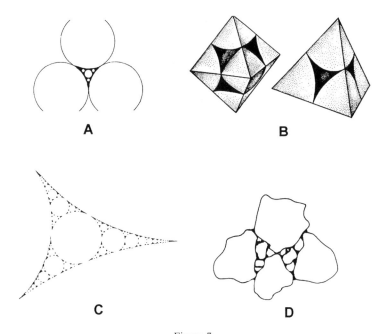

Figure 7
Schematic models for dense packings. A: Circular disk model. B: Sphere model, left octahedral interstitial space, right tetrahedral interstitial space (modified after KESSLER, 1994). C: Appolonian packing (after MANDELBROT, 1991). D: Natural fractal packing.

complex geological structures and shapes, fractal methods have been applied to describe particle fragmentation processes, pore geometries and many other geopatterns (KAYE, 1989; XIE, 1993; KRUHL, 1993; TURCOTTE, 1997). Fractal packing models are not examined to such a great extent, although the fractal "Appolonian packing" is mentioned by a number of authors (PAPE et al., 1984; KAYE, 1989). Two-dimensional "circular disk models" and three-dimensional "sphere models" are both fractal although they are not called fractal by engineers. As opposed to the Fuller distribution, they give rise to discontinuous gap gradings with a strong dominance of the coarse first-order grain fraction. This trend is also visible in the cumulative distribution curves of the dense asphalt mixtures that were empirically developed. Obviously there are serious difficulties concerning the calculability of these packings. The estimation of the radii looks easy for the first, second and third order of grains, but becomes very complex towards smaller radii because the interstitial spaces become more complex as more and different radii are introduced into the calculation. The "Appolonian packing" is also quite complex since it is not scale-invariant and not self-similar which makes the quantification of its fractal dimension indefinite. This is in good accord with the results proposed by KESSLER (1994) who made the first attempt towards a three-dimensional packing model. Even though the particles are "Euclidean" both models illustrate the fractal packing concept of a finite volume or area that is filled by an infinite number of particles with an infinite surface area.

In practice it is necessary to prove whether the mixing process or the type of deposition of the investigated material is better reflected by a two-dimensional or a three-dimensional model. Polyphase materials, in which the pores between large grains are filled by inflush of smaller grains, can be compared to a sequence of two-dimensional filter planes and explained by a two-dimensional model. The diameters of the grains of following orders in polyphase deposits would rather respond to a two-dimensional "Appolonian packing" and be smaller than the actual pore space in three dimensions. Graded mixtures and mass deposits are better explained by three-dimensional models because the grains are allowed to move more or less freely during mixing and transport and the pore spaces can be occupied by ideally fitting grains in three dimensions.

Processing in the rock industry normally starts with fragmentation through blasting and grinding, followed by separation of the material into single grades and mixing from the stored single grades. In terms of grain size distributions, the material obviously has a fractal composition which from the beginning undergoes separation, and may be remixed to a fractal grain size distribution again, for example to produce the Fuller distribution. In fact the initial fragmented material as such is suitable for many technical uses, for example in stable foundations. However, qualified mineral mixtures normally undergo the separation and mixing processes which appears surprising, especially for the fractal mixtures. The initial fractal composition could possibly be better used also for high quality mixtures under the assumption that the other technical demands are fullfilled, making the processing of such mixtures considerably more economical.

In theory the relative density of a fractal mix approaches the mineral density because the porosity diminishes to zero. Consequently the shear strength of a fractal rock theoretically increases towards the value of the mineral strength. During the shear-box experiments on fractal gradings, the machine reached its limits for the shear strength at a confining pressure of 0.75 MPa although it is designed for confining pressures reaching 1.2 MPa. In practice, densely packed grain supported conglomerates are known to be mechanically very resistant and sufficiently impermeable to act as barriers for fluids. For a better calculation of mechanical parameters of such rocks it might be useful to analyze the type and quality, e.g., the fractal density of the packing in addition to the simple bulk density. Furthermore, the mechanical properties are strongly controlled by grain shape, roughness and surface area, which are also fractal but scale-dependent in many natural examples. In addition to the pore volume used in many calculations, length or area values are needed that characterize grain geometries at the scale relevant to particular engineering problems. Surprisingly, fractal methods and concepts are still hesitatingly being applied for such purposes. In order to deduce physical and mechanical rock properties from packing densities and grain geometries, further investigations are needed of combined three-dimensional fractal techniques in engineering geology.

Conclusions

1. Dense materials show extraordinary mechanical properties and are very attractive scientifically and economically. The infinite fractal packing concept applies best for dense packings in which the highest possible strength and the lowest possible permeability are present or required.

2. A high initial fractality of grains and gradings is fundamental in natural and technical rock fragmentation. Fractal compositions and high fractal dimensions of irregular shaped particles are destroyed by high selective transport energies that, in the extreme, produce single graded rocks composed of rounded "Euclidean" grains.

3. Fractal models for dense packings have very complex space geometries as a result of their multisize grain composition. The basic processes of natural sedimentation, or material mixing, can be reflected either by two-dimensional "circular disk models" or by three-dimensional "sphere models." The "Appolonian-packing model" is the most universal circular disk model.

4. A thorough understanding of fragmentation processes, packing densities, single grain geometries and grain surface conditions is of great economic value in the rock industry. Fractal models and fractal material testing methods give numbers to complex structures and irregular shapes. As such, they are the key to a better calculation of geomechanical and geotechnical rock properties.

Acknowledgements

I am very grateful to H. J. Pape for his helpful suggestions and especially for the mathematical derivation of the frequency size distribution and the cumulative distribution. I also wish to thank the second referee who enhanced the manuscript with his critical comments.

REFERENCES

BAMFORD, W. E. (1978), *Suggested Methods for the Quantitative Description of Discontinuities in Rock Masses*, Intern. J. Rock Mech. Min. Sci. and Geomech. Abstr. *15*, 319–369.

BOUMA, A. H., *Sedimentology of Some Flysch Deposits. A Graphic Approach to Facies Interpretation* (Elsevier, Amsterdam 1962).

FULLER, and THOMPSON (1907), *The Laws of Proportioning Concrete*, Trans. Am. Soc. Eng. *59*, 67.

KAYE, B. H., *A Random Walk Through Fractal Dimensions* (VCH-Verlagsgesellschaft, Heidelberg 1989).

KESSLER, H-G. (1994), *Kugelmodelle für Ausfallkörnungen dichter Betone (Sphere Models for Gap Gradings of Dense Concretes)*, Betonwerk und Fertigteiltechnik *11*, 63–76.

KRUHL, J. H., *Fractals and Dynamic Systems Geoscience* (Springer, Berlin 1993).

KUENEN, PH. H. (1964), *Deep sea sands and ancient turbidites*. In *Turbidites. Developments in Sedimentology 3* (Elsevier Amsterdam 1964) pp. 3–33.

LEE, Y. H., CARR, J. R., BARS, D. J., and HAAS, C. J. (1990), *The Fractal Dimension as a Measure of the Roughness of Rock Discontinuity Profiles*, Int. J. Rock Mech. Min. Sci. and Geomech. Abstr. *27*, 453–464.

MANDELBROT, B. B., *Die fraktale Geometrie der Natur* (Birkhäuser, Berlin 1991).
ORTIGAO, J. A. R., *Soil Mechanics in the Light of Critical State Theories* (Balkema, Rotterdam 1995).
PAPE, H., RIEPE, L., and SCHOPPER, J. R. (1984), *The Role of Fractal Quantities as Specific Surfaces and Tortuosities for Physical Properties of Porous Media*, Part. Character. *1*, 66–73.
PAPE, H., and SCHOPPER, J. R. (1987), *Stereological Determination of Grain Size Distribution in Consolidated Aggregates on the Base of the Fractal Concept*, Acta Stereologica *6/III*, 827–832.
PRINZ, H., *Abriß der Ingenieurgeologie* (Ferdinand Enke Verlag Stuttgart 1991).
RICHARDSON, L. F. (1961), *The Problem of Contiguity: An Appendix of Statistics of Deadly Quarrels*, General Systems Yearbook *6*, 139–188.
ROSIN, P., RAMMLER, E., and SPERLING, K. (1933), *Korngrößenprobleme des Kohlenstaubs und ihre Bedeutung für die Vermahlung*, Berichte der Technisch-Wirtschaftlichen Sachverständigenausschüsse des Reichskohlenrats, C52, 1–25.
STEFFEN, H. (1986), *Abdichtungssysteme für Deponien*, Veröffentlichungen Grundbauinstitut, LGA Bayern *47*, 33–82.
SUTEANU, C., IOANA, C., MUNTEANU, F., and ZUGRAVESCU, D. (1993), *Fractal, Aspects in Solids Fragmentation; Experiment and Model with Implications in Geodynamics*, Revue Romaine de Geophysique *37*, 61–78.
TAKAYASU, H., *Fractals in the Physical Sciences* (Manchester University Press 1989).
TURCOTTE, D. L., *Fractals and Chaos in Geology and Geophysics* (University Press Cambridge 1997).
WASSON, R. J. (1977), *Last-glacial Alluvial Fan Sedimentation in the Lower Derwent Valley, Tasmania*, Sedimentology *24*, 781–799.
XIE, H., *Fractals in Rock Mechanics* (Balkema, Rotterdam 1993).
ZTV-ASPHALT-StB 94, *Zusätzliche, technische Vertragsbedingungen und Richtlinien für den Bau von Fahrbahndecken aus Asphalt* (Forschungsgesellschaft für Straßen und Verkehrswesen, 799, 1994).

(Received April 7, 1998, revised January 18, 1999, accepted January 21, 1999)

To access this journal online:
http://www.birkhauser.ch

Fractal Characterization of Particle Size Distributions in Chromitites from the Great Dyke, Zimbabwe

TOM G. BLENKINSOP[1] and T. R. C. FERNANDES[2]

Abstract—Chromitite from seams in the early Proterozoic Great Dyke, Zimbabwe, has three types of microstructure. Grains in intact samples have an average of just over five slightly curved grain boundaries around each polygonal grain, and triangular-shaped triple grain junctions, some with grain boundaries intersecting at 120°. These features show adjustment to a minimum surface energy configuration. Samples with extension microcracks have smaller particles on average, which are more inequant and have a stronger preferred orientation than particles in the intact samples, due to fragmentation by impingement microcracking. Microfaults have still smaller average particle sizes, but more equant and less well orientated angular fragments, formed by sliding and rotation of particles after linkage between extension microcracks. Intact samples have a curved relationship on a log-log plot between cumulative numbers of particles and grain size. This particle size distribution evolves with strain to a linear, fractal relationship in the microfaults, with a fractal dimension of 2.8. The changes in particle size distribution are consistent with constrained comminution, and an additional process of selective fracture of larger particles. The degree of cataclasis is an important factor in determining chromite ore quality.

Key words: Cataclasis, particle size distribution, fractal, microfault, chromite.

Introduction

Chromitite from seams in the Great Dyke is a major mineral resource in Zimbabwe, used for the production of ferrochromium alloys. However, there is considerable variation in the quality of the ore, even within a single seam. The quality is related to primary characteristics of the ore, and to secondary processes, which have affected the physical and mineralogical properties of the chromite (FERNANDES, 1987, 1997). One of the most important secondary processes is the degree of cataclastic deformation, which deteriorates ore quality. The aim of this study is to describe and understand the cataclastic processes that have affected the ore, as a first step towards predicting ore quality from cataclastic deformation features.

Cataclasis is fracture, sliding and rotation of rigid particles which creates a characteristic microstructure of angular to sub-angular fragments with a range of

---

[1] Department of Geology, [2] Institute of Mining Research, University of Zimbabwe, P.O. Box MP 167, Mount Pleasant, Harare, Zimbabwe.

sizes. The distribution of fragment sizes, or particle size distribution (PSD), reveals useful information about cataclastic deformation mechanisms (e.g., BLENKINSOP, 1991). PSDs of cataclastic rocks are commonly fractal, and can be described by the relation between the number of particles with a size greater than S, ($N > S$), and S:

$$N > S \sim S^{-D}$$

where D is the fractal dimension. Experiments, experience from industrial crushing and milling, and studies of naturally deformed rocks, show that D depends on rock type, energy input, strain and confining pressure, as well as the cataclastic process (BLENKINSOP, 1991). For example, extension microfracture in granitic rocks is characterized by D in the range 2.0 to 2.5, whereas shear microfracture has D values of 2.6 or greater. A value of $D = 2.6$ is common in many natural and experimental gouges, and is predicted by the theory of constrained comminution (SAMMIS et al., 1987). "Constrained comminution" means that the probability of neighbouring particles having equal sizes is minimized by microfracture, resulting in lower tensile stress concentrations at grain contacts. PSDs of samples in this study were measured and compared to previous observations and theoretical predictions in order to understand the cataclastic processes in chromitites of the Great Dyke.

Cataclasis affects ore quality on the atomic scale by causing strain-induced cation migration, and a change from the perfect cubic symmetry expected in intact chromites to tetragonal symmetry, which is demonstrated by changes in mineral composition, crystallographic and magnetic properties of the deformed ores compared to the undeformed ores (FERNANDES, 1997, 1999; FERNANDES and LANGFORD, 1999). This study focuses on the cataclastic processes that have caused such changes at the scale of a thin section.

Geological Background and Sample Locations

The Great Dyke is a layered ultramafic-mafic intrusion over 500 km long and between 2 and 11 km wide that intruded through the centre of the Archaean Zimbabwe craton at 2.450 ± 0.015 Ga (KAMBER et al., 1996 recalculated after HAMILTON, 1977). Gravity studies show that the Dyke has a Y-shaped transverse section, with a stem in places at least 10 km deep (PODMORE, 1982, 1983; PODMORE and WILSON, 1987). The intrusion of the magma occurred in two or possibly three major magma chambers (the North and South Chambers, Fig. 1), which consist of a lower ultramafic sequence of dunites and bronzitites, and an upper mafic sequence of gabbros and norites (PRENDERGAST and WILSON, 1989). The chromitites occur in seams 10 to 100 cm thick within the ultramafic sequence, parallel to the layering of the Great Dyke. There are up to 11 seams, divided into an upper group of 3 and a lower group of 8, numbered from 11 at the base to 1 at the top (PRENDERGAST and WILSON, 1989).

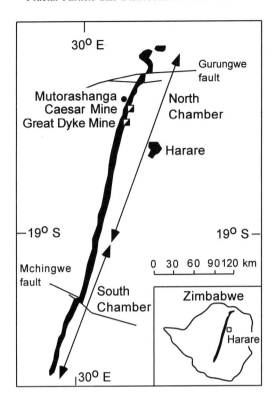

Figure 1
The Great Dyke in Zimbabwe, showing the North and South Chambers, the Gurungwe and Mchingwe faults, the location of Caesar and Great Dyke mines and Mutorashanga, where the samples in this study originate (see Table 1). Inset shows the location of the Great Dyke and Harare within Zimbabwe.

The whole length of the Great Dyke is cut by numerous faults on all scales, at a high angle to the trend of the dyke. East-west normal faults are the most prevalent. The sinuous trace of the Great Dyke at the north end of the craton is probably due to deformation in the Pan-African Zambezi belt, and dextral displacements on large faults, such as the Gurungwe fault (Fig. 1) may also be due to this deformation. Reverse faults and hinge faults are also common in the north. East-west to southeast-northwest faults cutting the southern part of the dyke have dextral separations up to several kilometres (e.g., Mchingwe fault, Fig. 1). The Great Dyke has therefore been affected by extensive discontinuous deformation, but the age of deformation and of the cataclasis described here is unclear.

The samples in this study come from Great Dyke and Caesar Mines in the North Chamber (Fig. 1, Table 1). Deformed samples from seam 4 were compared to undeformed samples from seam 7.

Table 1

Characteristics of chromite particles in the Great Dyke, Zimbabwe. Shape, resultant magnitude, and fractal dimension are defined in the text. Upper and Lower Limits are limiting values of the particle size used in the regression

	Intact (int)	Extension microfracture (ext)				Microfault (mf)	
Sample	26int	45Aext	45Bext	7ext	13Aext	54mf	7mf
Location	Mutorashanga	Gt. Dyke mine	Gt. Dyke mine	Caesar mine	Caesar mine	Gt. Dyke mine	Caesar mine
Seam	7	4	4	4	4	4	4
Magnification	49	84	84	84	168	84	168
Average size, mm $\times 10^2$	11.1	8.25	6.26	6.24	3.05	3.48	1.52
Maximum size, mm $\times 10^2$	25.3	68.7	32.0	38.1	20.0	57.7	16.8
Minimum size, mm $\times 10^2$	0.11	0.98	0.66	0.65	0.49	0.66	0.33
Shape	1.67	3.17	3.19	2.90	3.29	2.55	2.50
Resultant magnitude	0.61	0.70	0.72	0.73	0.79	0.64	0.67
Fractal dimension						2.761	2.878
Correlation coefficient						0.999	0.994
Standard error						0.0192	0.0378
Upper limit						0.1	0.07
Lower limit						0.02	0.01
Number of particles	184	613	600	506	430	748	501

Methods

Thin sections and polished blocks of the chromitites were examined under the optical microscope. PSDs were measured on photomicrographs of the samples at enlargements of $\times 49$, 84 or 168 (Table 1). Particle size was measured as the average of the maximum particle dimension and its perpendicular bisector. The average, maximum and minimum particle sizes, the average shape of the particles (defined as the ratio of the two axes), and the resultant magnitude of the direction of the long axes, were also measured. The resultant magnitude (R) of N particles is given by

$$R = 1/N \left(\sum \sin \theta^2 \Big/ \sum \cos \theta^2 \right)^{1/2}$$

where θ is the angle between the particle long axis and an arbitrary reference direction. R measures the degree of preferred orientation of the particles.

Linear regression of $\log N > S$ against $\log S$ was used to find the fractal dimension of those samples with a fractal relationship, which is given in Table 1 with the correlation coefficient and standard error of regression. The two-dimensional fractal dimensions were increased by 1 to obtain the three-dimensional fractal dimension (cf. SAMMIS et al., 1987).

Results

The samples can be grouped into three types of microstructure.

Intact Microstructure

Samples with little visible deformation have a distinctive microstructure of approximately equal-sized polyhedra of single crystals of chromite. The average size of sample 26int is 0.1 mm (Table 1). The edges of the polyhedra are commonly slightly curved, and several of them intersect at approximately 120° at grain triple junctions (Fig. 2). Small interstices of silicate exist at most grain triple junctions and along many grain boundaries (Fig. 2). These silicates are probably serpentine after

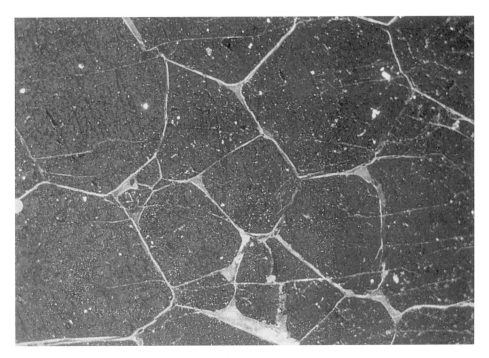

Figure 2
Intact microstructure (sample 26int, polished thin section. Width of field of view 2.83 mm). Polyhedral grains have from 3 to 6 slightly curved edges, several of which intersect in grain triple junctions at approximately 120°. Light material in grain triple junctions and along grain boundaries is a silicate, probably serpentine after olivine.

primary olivine. Figure 3 displays outlines of individual polyhedra of sample 26int, from which the number of sides around each polyhedra have been counted. The number of sides on each grain is shown as a histogram in Figure 4. Three to

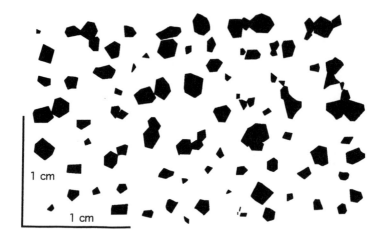

Figure 3
Tracings of individual chromite grains of sample 26int from polished thin sections. These grains are in contact in the sample, but have been partly separated for greater clarity in the drawing.

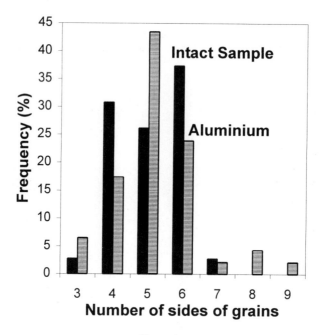

Figure 4
Histogram of a number of grain edges/grain in sections of sample 26int compared to annealed aluminium from SMITH (1964).

six-sided polyhedra are common, and the average number of grain sides is 5.065. The average shape of grains in the primary microstructure (Fig. 5), and the resultant magnitude (Fig. 6), are the lowest of all microstructures. The PSD for the intact sample is shown in Figure 7. It is clearly curved and does not exhibit any straight line portion.

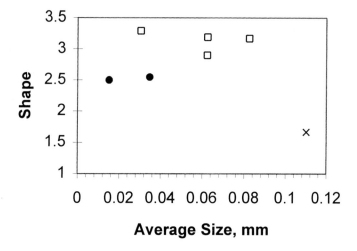

Figure 5
Shape of particles (ratio of long to short axes) against average size. × — Intact microstructure, □ — Extension microcrack microstructure, ● — microfault.

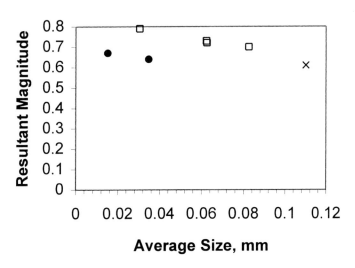

Figure 6
Resultant magnitude (see text for definition) against average size. Symbols as in Figure 5.

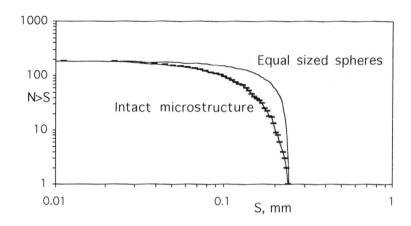

Figure 7
Number of particles with a diameter greater than S, ($N > S$), as a function of S for the intact microstructure (sample 26int) compared to the theoretical distribution of equal-sized spheres.

Extension Microcrack Microstructure

Extension microcracks are ubiquitous in all deformed samples. They are generally intragranular and connect contact points between grains or points where grain boundaries approach each other closest (Fig. 8), avoiding pores at triple grain junctions. The dominant systematic extension microfractures are straight and have a strong preferred orientation. Original grain boundaries are clearly visible in this microstructure (Fig. 8), showing the same characteristics as the primary microstructure described above. The purely extensional mode of the microcracks can be deduced from the lack of shear displacement on grain boundaries. The average grain size of the samples with the extension microcracks is distinctly lower than the intact microstructure (0.03 to 0.08 mm), but the shape and resultant magnitudes of the extension microcrack microstructure are higher (Table 1, sample numbers with "ext" suffixes, Figs. 5 and 6). The PSD of samples with extension microcracks are shown in Figure 9; they are distinctly less curved than the intact sample, but the relation is not linear and does not justify calculating a fractal dimension.

Microfaults

Microfaults have a completely different microstructure of randomly orientated angular fragments with widely ranging sizes (Fig. 10). Aggregates of cemented grain fragments demonstrate that microfaulting was a repeated process. No relic of the intact microstructure is visible. The average particle size in the

Figure 8
Extension microfracture microstructure (sample 45Aext, polished thin section. Width of field of view 2.83 mm). Microcracks generally link points of contact between adjacent grains and avoid pores at grain triple junctions. The dominant systematic microcracks have a strong preferred orientation. Original polygonal grain boundaries are visible, and lack of shear displacement of the boundaries shows the extensional mode of the microcracks.

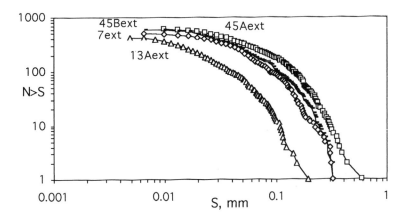

Figure 9
Number of particles with a diameter greater than S, $(N > S)$, as a function of S for the extension microcrack microstructure.

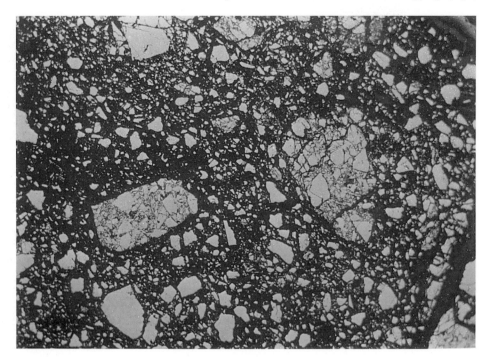

Figure 10
Microfault (sample 54mf, polished thin section. Width of field of view 2.83 mm). Angular fragments have a large range of sizes. An aggregate of cemented fragments is seen near the centre. Dark areas around the visible fragments are chromite which has not taken a polish.

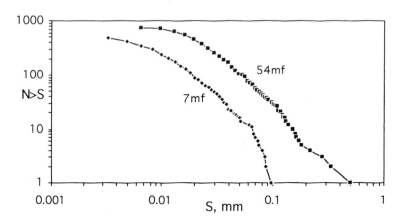

Figure 11
Number of particles with a diameter greater than S, ($N > S$), as a function of S for the microfaults.

microfaults is lower than the other microstructures (Table 1, sample numbers with "mf" suffixes). The shapes and resultant magnitudes of fragments in the microfaults are lower than in the extension microfracture microstructure, but higher than in the intact microstructure (Figs. 5 and 6). The particle size distributions can be divided into three parts: a curved part at low S values, a linear part at intermediate values, and an irregular part at highest S values (Fig. 11). Regression of the PSD over the visually estimated linear part of the data yields fractal dimensions of 2.76 and 2.88 (Table 1).

Discussion

The three microstructural types form a progressive sequence with increasing deformation from the intact microstructure to extension microcracks and finally the microfault microstructure. Some aspects of this evolution are particularly well demonstrated in the chromitites because they have a relatively simple initial microstructure, as described below.

The intact microstructure is an excellent example of a microstructure adjusted to achieve minimum surface energy or foam structure. The shape with the minimum surface energy that tessellates in three dimensions is the truncated octahedron, consisting of eight hexagonal faces and six square faces (THOMPSON, 1887). However, this shape does not satisfy isotropic surface energy requirements because there are unequal angles between grain edges at the corners of the square faces where some pairs of grain edges meet at 90°, and others at 120°. An isotropic distribution of surface energy can be achieved by curvature of the grain edges and faces to allow the grain edges to meet at the ideal angle of 109° (SMITH, 1964). This microstructure has been described from metals (SMITH, 1964), and appears to be similar to the microstructure of the intact chromitites. To further confirm the similarity, the number of sides/grain in a section of equilibriated microstructure in aluminium (illustrated in SMITH, 1964) is compared to the intact chromitite in Figure 4. The similarity is striking, and confirmed by the similarity in the average number of sides/grain, which is 5.196 in the aluminium (46 grains, sample standard deviation 1.222) and 5.065 in the chromitites (107 grains, sample standard deviation 0.954). The truncated octahedron has an average of 5.143 sides on each face, but this is not expected to correspond to the above values because they were measured on arbitrary two-dimensional sections which are not generally the faces of the octahedra.

The rounded grain corners observed at triple grain junctions in sections are also evidence for adjustment to surface energy equilibrium. The weak shape fabric and low degree of grain shape orientation in the intact microstructure may result from microcracks which are present even in the least deformed sample. The PSDs of the intact samples are strongly curved, and clearly non-fractal. For comparison, the theoretical distribution of sections through N equal-sized spheres with a diameter of D, is (e.g., EXNER, 1972):

$$N > S = N(1 - (S/D)^2)^{1/2}.$$

This curve is shown in Figure 7 for 184 spheres with the diameter equal to the maximum size of grains in the intact sample, 26int. This is an end member PSD, in which the curvature is purely due to the section effect on equal-sized particles. The intact sample PSD is also curved, but differs from the theoretical distribution of equal-sized particles because the chromites have a finite, but limited, range of sizes in three dimensions. The difference between the two curves is not due to "roll-off" (see below) because it occurs at all particle sizes, whereas roll-off occurs preferentially at low particle sizes.

Since the chromitite layers of the Dyke are considered to have formed either by gravitational settling or direct crystallization from a melt, these presumably euhedral magmatic chromites must have been annealed to produce the equilibrium microstructure. This probably occurred during subsolidus cooling of the Dyke, since the foam texture is ubiquitous and no subsequent heating event is likely to have affected the Dyke uniformly.

Impingement microcracks form due to tensile stress concentrations at contact points between grains. Theoretical elastic stress distributions around indenters demonstrate that impingement microcracks radiate from contact points (Fig. 12). Several distinctive features demonstrate that the extension microcracks in the chromitites formed by impingement between grains. The geometry of the microcracks, linking between contact points, is the most diagnostic feature. Microcracks which do not appear to link contact points may in fact do so out of the plane of section: this is suggested by the observation that their tips are near the closest approach of grain edges. The conspicuous lack of microcracks around grain triple junctions confirms the impingement mechanism, which is also consistent with the predominantly intragranular nature of the microcracks. Impingement microcracks will not form in a zero porosity aggregate unless grain boundaries fail: the role of the intragranular silicates may be important in this respect. Elastic strain contrasts across grain boundaries due to any elastic anisotropy could also contribute to microfracturing.

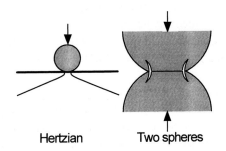

Figure 12
Microfractures predicted by stress analysis of the indentation of a sphere on a plane (Hertzian configuration, after LAWN and WILSHAW, 1975) and loading of two spheres (WONG, 1990). Microfractures radiate from the contacts between the two surfaces, similarly to the geometries seen in the extension microcrack microstructure.

The significance of the lower average particle sizes in the extension microcrack microstructures is not immediately obvious since the original grain size of the samples is unknown. However, the maximum particle size of the samples with extension microcracks is similar to, or larger than, the intact sample, suggesting that the original average size also must have been similar or larger. Therefore the decrease in average particle size is due to the subdivision of grains by microcracking.

Impingement microcracks commonly have a consistent orientation on the scale of a thin section and even regionally (e.g., LESPINASSE and PECHER, 1986), because they form perpendicular to the least principal stress (σ_3). Impingement geometries of grains modify the orientation of the regional σ_3 on a microscopic scale, resulting in a dispersion of microcrack orientations about the regional σ_3. The increase in the value of the resultant magnitude of particles from the intact to the extension microcrack microstructure is due to the increase in orientated extension microcracks, which determine the shape of many grain fragments. The high value of the resultant magnitude confirms the visual impression that individual microcracks do not substantially deviate from σ_3. This can be attributed to the coherent, tessellating and low porosity microstructure of the intact rocks, which creates a relatively homogeneous stress field. The PSDs of the extension microcrack microstructure are distinctly less curved than those of the intact microstructure, but cannot be described as fractal. This contrasts with PSDs of samples with extension microfractures reported by BLENKINSOP (1991), which are fractal. The difference can be attributed to the strongly curved, non-fractal PSD of the intact chromitites, which requires more cataclastic strain to evolve towards a fractal distribution.

The growth of faults from linkage of extension microcracks is a well-known phenomenon from experimental studies (e.g., PENG and JOHNSON, 1972) and studies of naturally deformed rocks (e.g., BLENKINSOP and RUTTER, 1986). This process is also seen in the chromitites of this study, in samples where incipient microfaults can be seen to have formed from linked extension microcracks. Grain boundary microcracks played a role in microcrack linkage (cf. MENÉNDEZ et al., 1996). Direct interaction of microcrack stress fields was also a linkage mechanism, as shown by microcrack tips that deviate in the vicinity of other microcracks (cf. KRANTZ, 1979).

The randomly orientated angular fragments in the microfaults formed by the sliding and rotation of particles, as revealed by the disappearance of all traces of the grain boundaries that are seen in the intact microstructure. The average grain size of the particles in the microfaults is generally lower than the extension microcrack and intact samples, showing that comminution of fragments continued in the microfault matrix. Reductions in the shape and resultant magnitude of the particles from the values of these parameters in the extension microcrack microstructure reflect the increasing importance of fragment attrition and rotation.

The PSDs of particles in the microfaults have a clear linear portion which allow fractal dimensions to be calculated. The departure of the PSD from the straight line at low values of S is expected because small particles are undersampled: this "roll-off" is well known from fractal studies (e.g., WALSH et al., 1991; PICKERING et al., 1995; BLENKINSOP, 1995). Further studies with the scanning electron microscope are planned to extend the range of particle sizes examined to lower values. The irregular part of the PSD at large values of S is also expected because there are insufficient measurements of the largest particles to define the PSD accurately.

The fractal dimension of particles in microfaults is close to the value of 2.58 predicted by the theory of constrained comminution (SAMMIS et al., 1987), and observed in many natural and experimental fault gouges (e.g., SAMMIS et al., 1986; SAMMIS and BIEGEL, 1989; MARONE and SCHOLZ, 1989). However, the values reported in this study (2.76 and 2.88, Table 1) are slightly but significantly higher than 2.58 (the standard errors are 0.02 and 0.04), suggesting that a further process of particle size reduction occurred in which the larger particles were selectively broken down. This process has also been noted in a previous study of cataclasites (BLENKINSOP, 1991).

The PSDs and microstructures therefore show a progressive evolution with strain, from the intact sample that has a curved PSD to a linear, fractal PSD in the microfaults. The grain size decreased during this evolution, as displayed on Figure 13, where PSDs from one example of each stage are shown, normalized

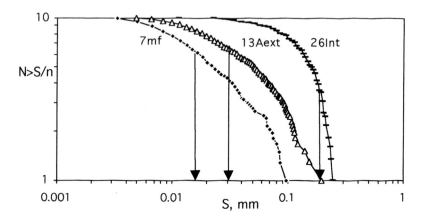

Figure 13
Normalized number of particles with a diameter greater than S, $N > S/n$, as a function of S from the three types of microstructure. Cumulative numbers of particles have been normalized to the total number measured (n) to facilitate comparison. The average particle size is shown by the arrow and decreases from the intact microstructure (26int) to the extension microcrack microstructure (13Aext) to the microfault (7mf). The PSDs evolve from highly curved to fractal.

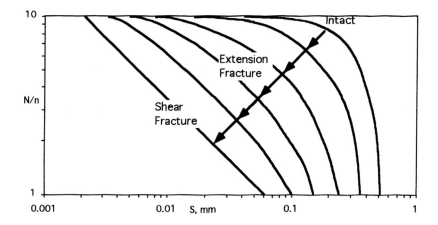

Figure 14
Schematic evolution of PSDs from intact samples to microfaults, showing the same features as Figure 13 schematically.

to the number of grains measured in each sample. The evolution is delineated schematically with several additional intermediate stages in Figure 14.

Conclusions

Cataclasis in the chromitites of the Great Dyke manifests typical features of natural and experimental cataclasis of low porosity, crystalline materials. The cataclastic evolution is particularly clear because of the regular microstructure of the intact chromitites, which is a nearly perfect example of a foam texture adjusted to achieve minimum surface energy. Extension microcracks form by impingement, but have a strong preferred orientation due to the homogeneous stress field in the intact microstructure. Microfaults form by linkage of microcracks. The particle size distributions evolve from a curved distribution in the intact microstructure, to a linear, fractal distribution in the microfaults. Constrained comminution occurred during this evolution, and an enigmatic additional process of selective microfracture of larger particles increased the fractal dimension beyond the value for constrained comminution.

This study has identified cataclastic processes in the chromitites of the Great Dyke at the scale of a thin section. This is the first step towards understanding the role of cataclasis in affecting ore quality. Fractal dimension may be a useful way of quantifying cataclasis in the light of previous studies that establish that fractal dimension evolves with strain. Some method of assessing the distribution of the microstructures on a larger scale will be necessary to extend the study, because the different types of cataclastic microstructure can occur in close association, even together in a single thin section.

Acknowledgements

We are grateful to E. Rutter and J. Kruhl for useful reviews. This research was supported by the Commission on Tectonics (COMTEC) of the International Union of Geological Sciences (IUGS).

REFERENCES

BLENKINSOP, T. G. (1991), *Cataclasis and Processes of Particle Size Reduction*, Pure appl. geophys. *136*, 1–33.

BLENKINSOP, T. G. (1995), *Fractal measures for size and spatial distributions of gold mines: Economic applications*. In *Sub-Saharan Economic Geology* (eds. Blenkinsop, T. G., and Tromp, P. L.) Spec. Publs Geol. Soc. Zimbabwe *3*, pp. 177–186.

BLENKINSOP, T. G., and RUTTER, E. H. (1986), *Cataclastic Deformation of Quartzite in the Moine Thrust Zone*, J. Struct. Geol. *8*, 669–681.

EXNER, H. E. (1972), *Analysis of Grain- and Particle-size Distributions in Metallic Materials*, Int. Met. Rev. *159*, 25–42.

FERNANDES, T. R. C. (1987), *The Mineralogy of Chromites from Zimbabwe and their Metallurgical Characteristics*, M. Phil. Thesis, Univ. Zimbabwe.

FERNANDES, T. R. C. (1997), *Mineralogical Studies to Distinguish Two Types of Chromite from Zimbabwe*, D. Phil. Thesis, Univ. Zimbabwe.

FERNANDES, T. R. C. (1999), *Significance of Ferrimagnetism in Chromitites of the Great Dyke, Zimbabwe*, J. Af. Earth Sci. *28*, 337–348.

FERNANDES, T. R. C., and LANGFORD, J. I. (1999), *Line Broadening in Chromites from Zimbabwe Using High Resolution X-ray Diffraction*, Power Diffraction *12*, 93–105.

HAMILTON, J. (1977), *Sr Isotope and Trace Element Studies of the Great Dyke and the Bushveld Mafic Phase and their Relation to Early Proterozoic Magma Genesis in Southern Africa*, J. Petrol. *18*, 24–52.

KAMBER, B. S., WIJBRANS, J. R., BIINO, G. G., VILLA, I. M., and DAVIES, G. (1996), *Archaean Granulites of the Limpopo Belt, Zimbabwe: One Slow Exhumation or Two Rapid Events?*, Tectonics *15*, 1414–1430.

KRANTZ, R. L. (1979), *Crack-crack and Crack-pore Interactions in Stressed Granite*, Int. J. Rock Mech. and Mining Sci. and Geomech. Abstr. *16*, 37–47.

LESPINASSE, M., and PECHER, A. (1986), *Microfracturing and Regional Stress Field: A Study of the Preferred Orientations of Fluid-inclusion Planes in a Granite from the Massif Central, France*, J. Struct. Geol. *8*, 169–180.

LAWN, B. R., and WILSHAW, T. R. (1975), *Review Indentation Fracture: Principles and Applications*, J. Mater. Sci. *10*, 1049–1081.

MARONE, C., and SCHOLZ, C. H. (1989), *Particle-size Distribution and Microstructures within Simulated Fault Gouge*, J. Struct. Geol. *11*, 799–814.

MENÉNDEZ, B., ZHU, W., and WONG, T.-F. (1996), *Micromechanics of Brittle Faulting and Cataclastic Flow in Berea Sandstone*, J. Struct. Geol. *18*, 1–16.

PENG, S., and JOHNSON, A. M. (1972), *Crack Growth and Faulting in a Cylindrical Sample of Chelmsford Granite*, Int. J. Rock Mech. and Mining Sci. *9*, 37–86.

PICKERING, G., BULL, J. M., and SANDERSON, D. J. (1995), *Sampling Power-law Distributions*, Tectonophysics *248*, 1–20.

PODMORE, F. (1982), *The First Bouguer Anomaly Map of Zimbabwe*, Trans. Geol. Soc. S. Afr. *85*, 127–133.

PODMORE, F. (1983), *A Gravity Study of the Great Dyke, Zimbabwe*, D. Phil. Thesis, Univ. Zimbabwe.

PODMORE, F., and WILSON, A. H. (1987), *A reappraisal of the structure, geology and emplacement of the Great Dyke, Zimbabwe.* In *Mafic Dyke Swarms* (eds. Halls, H. C., and Fahrig, W. F.) Spec. Pap. Geol. Ass. Can. *34*, pp. 433–444.

Prendergast, M. D., and Wilson, A. H. *The Great Dyke of Zimbabwe—II: Mineralization and mineral deposits.* In *Magmatic Sulphides—the Zimbabwe Volume* (eds. Prendergast, M. D., and Jones, M. J.) (Institution of Mining and Metallurgy, London 1989) pp. 21–42.

SAMMIS, C. G., and BIEGEL, R. L. (1989), *Fractals, Fault-gouge and Friction,* Pure appl. geophys. *131*, 254–271.

SAMMIS, C. G., OSBORNE, R. H., ANDERSON, J. L., BADERT, M., and WHITE, P. (1986), *Self-similar Cataclasis in the Formation of Fault Gouge,* Pure appl. geophys. *124*, 53–78.

SAMMIS, C., KING, G., and BEIGEL, R. (1987), *The Kinematics of Gouge Deformation,* Pure appl. geophys. *125*, 77–812.

SMITH, C. S. (1964), *Some Elementary Principles of Polycrystalline Microstructure,* Metall. Rev. *9*, 1–48.

THOMPSON, W. (Lord Kelvin), (1887), *On the Division of Space with Minimum Partitional Area,* Phil. Mag. *24*, Fifth Series, 503–513.

WALSH, J., WATTERSON, J., and YIELDING, G. (1991), *The Importance of Small-scale Faulting in Regional Extension,* Nature *351*, 391–393.

WONG, T.-F. (1990), *Mechanical compaction and the brittle-ductile transition in the porous sandstones.* In *Deformation Mechanisms, Rheology and Tectonics* (eds. Knipe, R. J., and Rutter, E. H.) Spec. Publs. Geol. Soc. Lond. *54*, 111–122.

(Received April 11, 1998, revised/accepted November 7, 1998)

To access this journal online:
http://www.birkhauser.ch

Dynamics and Scaling Characteristics of Shear Crack Propagation

VADIM V. SILBERSCHMIDT[1]

Abstract—A model for a description of a shear crack (fault) propagation is proposed on the basis of the unification of continuum damage and fracture mechanics with ideas of fracture models for random media. Energy release linked with local failures of elements as a result of evolution of fracture processes at lower scale levels (stimulated by the stress concentration in the vicinity of the fault) is treated as a seismicity source. $2d$ simulations are performed for analysis of the effect of the size of pre-existing fault on characteristic features of rupture evolution. Scaling (including multifractal) character of crack propagation and of respective energy release is shown.

Key words: Shear crack, shear damage, earthquakes, crack-damage interaction, scaling, crack propagation.

Introduction

Recent decades of investigation and modelling of earthquakes are marked by intensive utilisation of fracture approaches in a description of characteristic features of seismicity. These methods supplement—to a certain extent—stick-slip schemes, based mainly on the well-known BURRIDGE and KNOPOFF (1967) model. An interest in both approaches is easily understandable, to the extent that both fracture and friction play an important role in real earthquakes (GROLEAU *et al.*, 1997). The emphasis on the use of fracture models is explained, on the one hand, by the recent elaboration of new simulation techiques and respective achievements in the study of failure evolution in complex and stochastic media (see HERRMANN and ROUX, 1990 and references therein) and, on the other hand, by approved universality, reflected in scaling parameters, of the rupture process realisation at different scale levels.

Still, the question of applicability of this type of model to the simulation of earthquakes should be cleared. An elaboration of a totally adequate description of *all* features of seismicity within one approach today is hardly possible. This can be explained not only by a wide spectrum of structural levels participating in seismic events, but also by a variety of mechanisms influencing them (fracture and friction

---

[1] TU München, Lehrstuhl A für Mechanik, Boltzmannstr. 15, D-85748 Garching b. München, Germany, e-mail: silber@lam.mw.tu-muenchen.de

being probably the most important). As, regards tradition we deal with an 'output' of earthquakes in the form of seismic signals/energy release, thus we have no direct evidence of real mechanisms and their interaction which caused the event. Without direct information pertinent to the reasons for seismicity, it was natural to look for models based on a non-contradictory physical or mechanical presentation of an epicentre and producing results analogous to seismic data. Difficulties of direct comparison of a model and real data presupposed the search for generalised characteristics of seismicity, for universality class of phenomena. In this respect mechanical models of an epicentre can be useful for an analysis of partial mechanisms/features of seismicity.

Fracture models, used in earthquake simulations, can be roughly divided into two large groups: the first is based on the continuum mechanics description while the second mainly utilises procedures developed within statistical physics. The advantage of the first group is the possibility of exploiting both the spectrum of constitutive equations for media presentation and thoroughly elaborated procedures of fracture mechanics (LIEBOWITZ, 1968–1972). Still, these schemes also have considerable shortcomings: they usually deal with regular (mainly, smooth) geometrical objects as analogues of cracks/discontinuities, and consider the media under study to possess uniformly distributed properties. These disadvantages are overcome in procedures of the second group where both random distribution of properties and complex morphologies of fractures can be analysed. The price for these benefits is the utilisation of generally simplified concepts for the description of mechanical behaviour and transition to fracture under loading.

The idea of unification of both fracture and friction within single approaches was, and remains, an attractive aim for many investigators of seismic problems. In this paper we limit the short review part to recent models with a direct incorporation of elements of fracture analysis. The general background of fracture models of seismicity is an introduction of a pre-existing fault into a model description (e.g., CHEN et al., 1997). The seismic zone is presented as an elastic region with macroscopic inhomogeneities (faults) and randomly distributed fracture thresholds. A local shear fracture causes a stress redistribution as a result of a force imbalance. The general self-consistent procedure based on lattice Green functions can then be applied for calculation of this stress redistribution. The $2d$ model deals with blocks of a characteristic size of a few kilometres and a strike-slip fault embedded in the middle of the zone. Blocks neighbouring the fault experience frictional forces, preventing, up to a certain limit, displacements and thus building up overstresses in the vicinity of the fault. Earthquakes are related in this model to two types of fault instabilities: shear ruptures and/or slip. An analogous description, with an example of a $1d$ scale analysis, is developed by GROLEAU et al. (1997) on the basis of the model of XU et al. (1992) with utilisation of the coarse-graining for lattice Green's function.

A multi-level approach is proposed by SAHIMI and ARBABI (1996), in which failure at one scale is part of the damage/fracture process at a larger scale. In their $2d$ triangular network sites are connected with springs representing a rock portion at a lower scale level. Passing the threshold value for a length of the spring (which are randomly distributed) means its breakage and respective formation of microcracks. Another hierarchical description is developed by ALLÈGRE et al. (1995). Assuming a degenerated orientation of nucleated cracks, a relatively simple criterion for a fracture transition to upper scales is considered for the case of a $2d$-discretisation into square elements, namely, an alignment of three failed squares of the lower scale. Here an energy-form criterion is exploited for crack nucleation: its probability during the given time interval is proportional to the energy excess.

A thorough study of the interconnection of earthquakes and faults' development is fulfilled by SCHOLZ et al. (1991, 1993), SCHOLZ (1995, 1997). In this approach faults are treated as shear cracks with friction acting on their walls. Limitations of purely elastic crack models presuppose a transition to an elastoplastic fracture mechanic approach of the Dugdale-Barenblatt type accounting for inelastic deformations near the crack tip. A frictional breakdown zone of a finite length around the perimeter of the fault surface is introduced. The tip propagation is related to the overcoming of the shear strength of the surrounding rock.

In the author's recent paper (SILBERSCHMIDT, 1996) a shear damage accumulation was used for the description of the fracture process evolution at lower scale levels. Introduction of the critical damage value as a local failure criterion for elements of $2d$-discretisation allows a simulation of a spatio-temporal evolution of fracture process and consequent energy release. This scheme was applied to the modelling of seismicity and respective scaling characteristics were obtained. The proposed approach interpreted a fault as a result of the clusterisation of the locally failed zones. This paper spreads the description to cases with pre-existing faults, embedded into rock massif, accountable for the interaction between faults and damage accumulation.

Mechanisms of Crack-damage Interaction

Fracture evolution as a reason for seismicity in the Earth's crust is a considerably more complicated process as traditional failure problems linked with construction materials. This is explained by several factors: the presence of additional scales in fracture analysis (the most drastic manifestation is the difference in crack dimensions—extending some metres in specimens/structures and reaching hundred kilometres in the crust); extremely wide spectrum of characteristic times/rates of failure processes and different types of non-uniformness of properties at various scales. The problem of the formulation of a general description for such multilevel and highly random processes reflecting both slow processes of energy accumulation as a result of stresses' buildup under the action of tectonic forces and rapid release of it during earthquakes is rather complicated. One of the alternatives to such a description is the use of ideas developed within Continuum Damage Mechanics (CDM) (KACHANOV, 1986; LEMAITRE and CHABOCHE, 1990; KRAJCINOVIC, 1996). We note here that an ambiguity exists in the use of the term 'damage': within CDM it has a considerably different sense as in approaches of theoretical and statistical physics (CURTIN and SCHER, 1997; CURTIN, 1998; LYAKHOVSKY et al., 1997).

Applying CDM to the problem under study, we can limit our description to only the scale level of the main fracture(s) with account for the effect of the levels of lower scales by introduction of an additional variable—damage parameter. In contrast to the traditional form of CDM-parameters (see respective references in above-mentioned monographs), the damage parameter used in this paper has a deformational nature, it describes the part of strain linked with deformational manifestation (at the level under study) of nucleation, growth and coalescence of discontinuities of lower scales.

The decisive role of slip in seismicity serves as a reason for restraint of the description to a case of shear damage (SILBERSCHMIDT, 1993, 1996), considering for simplification the effect of other deformational modes on the process under study to be negligibly small. The respective kinetic equation for shear damage accumulation has the following form (SILBERSCHMIDT, 1994, 1998).

$$\frac{ds}{dt} = As^2 + Bs + D\langle \tau - \tau^* \rangle^n, \qquad (1)$$

where s is a damage parameter, A, B and D are material's parameters, an exponent n is close to unity (SILBERSCHMIDT, 1998), $\langle \ \rangle$ denote the Macauley's brackets (i.e., $\langle y \rangle = y$ if $y \geq 0$ and $\langle y \rangle = 0$ if $y < 0$). Generally speaking, s describes the part of the macroscopic shear deformation linked with the evolution of microshifts in media, thus parameters A, B and D should be obtained from experimental data on pure shear. An action of tectonic forces is represented in our model by shear stress τ. Friction forces and cohesion of the opposite sides of discontinuities prevent a realisation of slip before the stresses overcome respective threshold value τ^*. We note that the system described by (1) is characterised by the 'trigger-like' kinetics (SILBERSCHMIDT and CHABOCHE, 1994a): the process starts only after overcoming a potential barrier, linked with the presence of resisting forces (here, friction and cohesion).

The next step in modelling is an introduction of a fault into consideration. In order not to confront the description with complicated morphology of real faults, in the first approximation the pre-existing fault is considered to be a plane rectangle-form discontinuity, one side of which forms a common boundary with a rectangular cross section of a crust part under study. The massif to opposite sides of the fault is in a pure shear state induced by long-ranged tectonic forces, the action of which on the crack faces is characterised by shear stress τ^o. Instead of utilising a Dugdale-Barenblatt elastoplastic crack model with a frictional zone near the crack (fault) tip (COWIE and SCHOLZ, 1992; SCHOLZ, 1997), the ensemble of defects in the vicinity of the crack front, described in terms of CDM approach with damage parameter s, serves as an additional, irreversible mechanism in a process zone. For a given geometry/loading scheme the most probable direction of crack propagation is perpendicular to its front and it lies in the same plane as the pre-existing fault. Thereafter, the stress concentration in the vicinity of the crack tip can be described by the stress-intensity factor (SIF) of mode-II fracture K_{II} (LIEBOWITZ, 1968–1972)

$$\tau = \frac{K_{II}}{\sqrt{2\pi r}}, \tag{2}$$

where r is the distance to the tip. Concentration of stresses near the tip activates the processes of defects' development in this region, according to the damage accumulation law (1). A universal character of the stress decay in the vicinity of the crack tip allows the use of approximating functions (calculated either analytically or numerically) for a description of the change in the stress-intensity factor for a given geometry and loading type. These functions for different cases are given in various handbooks and can be applied to problems with similar geometry/loading. For the case under study the following approximation for the stress-intensity factor K_{II}:

$$K_{II} = 4.886\hat{l} - 11.383\hat{l}^2 + 28.198\hat{l}^3 - 38.563\hat{l}^4 + 20.555\hat{l}^5 \tag{3}$$

was taken from the respective handbook (MURAKAMI, 1987). Here $\hat{l} = l_f/L$ is a dimensionless length of the pre-existing fault; l_f and L are lengths of the fault and of the region under study containing this fault in a direction normal to the fault's front, respectively. These lengths thus correspond to the length of the crack in the epicentre and the size of the tectonic plate containing this fault.

Numerical Model: Account for Stochasticity

Non-uniformity of rocks and of realisation of various rupture mechanisms at different scale levels greatly influences characteristics of crack evolution. One of the possibilities to account for stochasticity in the model is the introduction of elements of discrete schemes into a continuous description. A set of models based on the lattice schemes was elaborated for simulation of fracture in disordered media (HERRMANN and ROUX, 1990). Unification of these concepts with CDM description was developed for mode-I damage and cracks (SILBERSCHMIDT and SILBERSCHMIDT, 1990; SILBERSCHMIDT, 1992; SILBERSCHMIDT and CHABOCHE, 1994b). A transition to the case of mode-II faults should leave the general ideas unchanged.

There are several means of introducing stochastic distribution of material properties into the model. One of the most intensively used ones is the distribution of stress threshold values; this concept is based on the simplest variant of strength theories. We prefer not to introduce non-uniformity directly in failure properties but to obtain them from the stochastic development of the rupture process evolution. Thus, in this model a shear modulus g is considered to be the only factor of stochasticity at the description level. The distribution is implemented in the course of discretisation of the cross section of the crust part under study into $\bar{M} = N_1 \times N_2$ rectangular elements. Each element characterised by a pair of indices (i, j) derives its shear modulus g^{ij} from the interval $[g_m, 2g_m]$, where g_m is the minimum value of modulus; details of the analogous distribution procedure are discussed elsewhere (SILBERSCHMIDT, 1996). The index i is a number of the elements' row perpendicular to the fault front, while j corresponds to rows parallel to it. One of the principle distinctions of elements in this model from those in approaches based on the Burridge-Knopoff model is that they are not the solid bodies connected by springs but the discretisation elements analogous, for instance, elements in finite elements method. Thus, all the features of continuum modelling remain, and discretisation into elements also provides additional possibilities for stochastic analysis.

The difference in crust response to the tectonic loading linked with the spatial distribution of its properties (shear modulus) results in the variation of shear stress distribution over elements. This can be described by means of the introduction of stress-concentration parameters K^{ij} for elements:

$$\tau^{ij} = K^{ij}\tau_0^{ij}, \qquad (4)$$

where τ_0^{ij} corresponds to shear stress in elements of the cross section under action of tectonic forces of the same magnitude, as in a case under study but for a virtual case of the fault's absence:

$$\tau_0^{ij} = \frac{\bar{M}g^{ij}\tau^o}{\sum_{i=1}^{N_1}\sum_{j=1}^{N_2} g^{ij}}. \qquad (5)$$

The damage accumulation equation (1) should be re-written for discretisation elements accounting for non-uniformity in the form:

$$\frac{ds^{ij}}{dt} = A(s^{ij})^2 + Bs^{ij} + D\langle \tau^{ij} - \tau^*\rangle^n. \qquad (6)$$

We note that material parameters, A, B and D also can be randomly distributed over the region under study in a general case of stochasticity. A relation for stress-concentration parameters in Eq. (4) has the following form:

$$K^{ij} = K_g^i K_f^{ij} \quad \text{(no summation with respect to } i\text{)} \qquad (7)$$

accounting for two mechanisms which cause shear stress change: stress concentration in the vicinity of the fault tip (described by K_f^{ij}) and the change of the common shear stiffness of the elements in the i-th row (K_g^i; discussed below). The second multiplier in Eq. (7) is obtained by the integration of Eq. (2) over elements of the i-th row:

$$K_f^{ij} = K_{II}^i \sqrt{\frac{\pi L}{N_2}} (\sqrt{j+1-n^i} - \sqrt{j-n^i}), \quad j > n^i, \tag{8}$$

where n^i is a number of elements, occupied by the crack. Equation (8) also means an overcoming within the proposed approach (based on the crust discretisation into elements) of the disadvantage of the concept of elastic crack with a non-physical singularity of stress distribution near its tip [according to Eq. (2)].

The mutual action of two factors: existence of the fault and spatial randomness in shear modulus distribution, results in a variation of shear stresses acting in discretisation elements [s. Eqs. (4) and (5)]. This, in turn, produces the difference in damage kinetics in elements according to Eq. (6). The local failure criterion for elements is introduced in the following form: the overcoming of a critical value s_c by the mode-II damage level for a given element results in its failure, release of accumulated energy and decrease of the element's shear modulus $g^{ij} \to \varepsilon g^{ij}$, $\varepsilon \ll 1$ (in calculations below $\varepsilon = 0$). Such decrease of the load-bearing capacity induces (by non-changing magnitude of tectonic forces) load redistribution in the cross section under study, which acts as an additional factor of stress concentration in non-failed elements. This is accounted for by the first multiplier in Eq. (7) which takes the following form:

$$K_g^i = \frac{\bar{G}^i}{\tilde{G}^i}, \tag{9}$$

where $\bar{G}^i = \sum_{j=n_0+1}^{N_2} g^{ij}$ is the initial shear modulus of the i-th row of elements, while $\tilde{G}^i$ being its current value; n_0 is a number of elements in each row occupied by the pre-existing fault.

Thus, elements with higher levels of stress-concentration parameters K^{ij} reach the critical damage level earlier than others. The local failure of the nearest neighbours of the elements occupied by the fault means its propagation in respective rows of elements. To the extent that such local events of failure/crack propagation occur at different places non-simultaneously, the fault's front after initial failures derives a tortuous form; it means also the transition from initial n_0 equal for all rows (for the initial relative fault length $\hat{l}_0 = n_0 L/N_2$) to current values of n_i and respective variation of the stress-intensity factor (SIF) along the front. This explains the necessity of introduction of K_{II}^i in Eq. (8); local SIF values are still described by approximation (3) although with respective relative lengths for each row of elements, normal to the crack front. The crack length variation along its front is a principle distinction of this approach from traditional crack models in fracture mechanics, in which crack fronts are considered to be smooth. In contrast to COWIE and SCHOLZ (1992), a crack propagation is linked here not simply with the level of shear stress that exceeds some critical value, this condition is reflected in shear damage accumulation law (6), but with overcoming a threshold level of damage parameter which is linked with the development of defects at lower scale levels, i.e., with transition to the next scale. The region with the fault-damage interaction presents here the process zone, developments in which determine specific features of the fault propagation.

Results of Simulations and Discussion

The above description is used as the basis for the elaboration of respective numerical algorithms. In simulations the cross section of the crust part under study, which contains the pre-existing fault, is discretised into 10,000 (100 × 100) elements. The initial relative length of the fault $\hat{l}_0$ was varied in simulations. Characteristic results of one of the simulations are shown in Figure 1. It is obvious that local failures occur initially only in the immediate vicinity of the fault's front, however

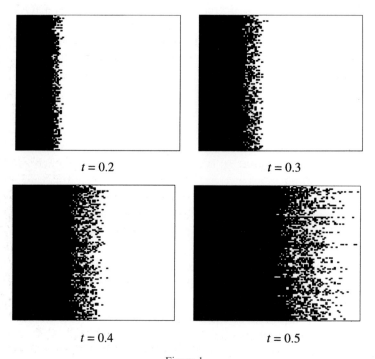

Figure 1
Propagation of a pre-existing fault and local failures of elements with time t (the elements belonging to the initial fault and failing at fulfillment of the local failure condition are shown in black; $\hat{l}_0 = 0.2$).

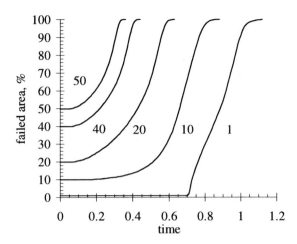

Figure 2
Effect of the length of the pre-existing fault (numbers near curves characterise the relative initial length in percent of the plate length L) on the dynamics of its propagation.

the crack propagation with time results in an increase of the zone with stress concentration sufficient for the initiation of elements' ruptures. Figure 2 presents results of numerical simulations for different lengths of the pre-existing fault. The shorter the length of the initial fault, the longer the period before the transition to a regime of a steady propagation of the crack. During this initial stage not only damage accumulates but also the overstresses buildup. The zone with occurring failures of elements corresponds to a transition of the fracture process from lower structural levels to the one under consideration; this region can be treated as a process zone (compare with SCHOLZ, 1997). For characterisation of its relative width we introduce two additional length parameters: l_{min} and l_{max}. The latter is the maximum length of the crack along its front (measured along the direction of its propagation from the side of the plate where the initial fault is), while the former is the minimal length. In other words, $l_{min} = \min_i l_i$, $l_{max} = \max_i l_i$, where $l_i = Ln_i/N_2$. Then the relative width of the process zone can be expressed as $(l_{max} - l_{min})/L$. It scales with a scaling exponent close to 4.0 with time (Fig. 3) at the initial stage of the fault propagation; the following deviation from such a behaviour is linked with the finiteness of the region being modelled.

The size of the pre-existing fault also effects temporal parameters of fault propagation. The characteristic onset of fault propagation linked with the initial local failures of elements in its front depends on the kinetics of damage accumulation. The latter, in its turn, is stipulated by the level of overstresses near the crack front. When the initial fault's length is greater than some characteristic level ($\hat{l}_0^* \approx 0.2$), it mostly does not influence the initial moment of crack propagation (Fig. 4). The end of the process, in contrast, non-monotonously decreases with an increase of the value of $\hat{l}_0$.

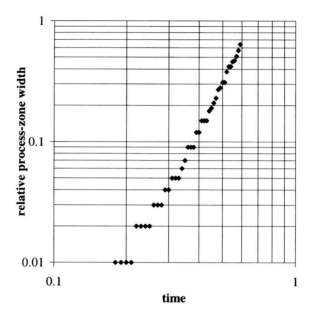

Figure 3
Evolution of the process zone: change of its relative width with time ($\hat{l}_0 = 0.1$, see comments in the text).

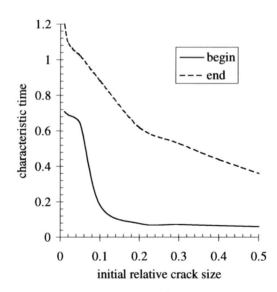

Figure 4
Effect of the initial crack length $\hat{l}_0$ on the characteristic time of fault propagation: moments of the first local element failure (begin) and of reaching the opposite side of the plate (end).

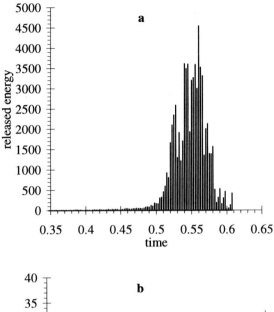

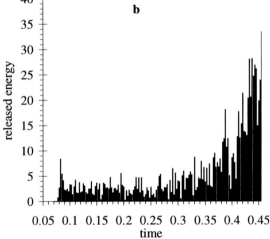

Figure 5
High- (a) and low-level (b) portions of energy release at fault propagation (one statistical realisation, $\hat{l}_0 = 0.2$).

Analysis of energy release is one of the main points in seismicity simulations. An energy release character during the fault propagation under the action of tectonic forces for the case under study is shown in Figure 5 for one statistical realisation of the stochastic distribution of stiffness g^{ij}. Though each statistical realisation is characterised by its own detailed character of energy release, the main properties of the process discussed further are the same. At the initial stage of relative rare and non-correlated events of element failure, the energy emission level is low and these events can be related to as foreshocks (Fig. 5b; compare the ordinate scale of both graphs in this figure). With the fault's growth and consequent buildup of overstresses, the process of local failures accelerates and the magnitude of released energy sharply increases. These results can be presented more traditionally for seismicity analysis form as diagrams of magnitude cumulative frequency dependence (Fig. 6). Here N is the number of seismic events with energy not higher than E. The use of such diagrams is linked with the well-known scaling character of seismicity, described by the so-called Gutenberg-Richter (1954) law:

$$\log N = a - bM, \qquad (10)$$

which links the number of events to its magnitude (moment) M. With the magnitude M being a logarithmic measure of the seismic energy E, one can obtain the scaling dependence of frequency and released energy in the form $N \propto E^{-\beta}$. Results of Figure 5 then can be interpreted as follows: there are two regions with different scaling behaviour, a low-energy part with the scaling exponent $\beta \approx 0.25$, and a high-energy part with a sharp decay of frequency; the scaling exponent for this portion is more than an order of magnitude larger than for the first one. The difference in scaling behaviour for these two parts exceeds results obtained for the seismicity in the crust region without the pre-existing fault (SILBERSCHMIDT, 1996) in which the scaling exponent for a low-energy part was $\beta \approx 0.4$. Such transition from one type of scaling to another with an increase in energy (magnitude) was reported by various authors for the analysis of earthquake catalogues (e.g., KAGAN, 1991). The level of value β (or b) is discussed in numerous publications, however the question of its universality is still open. Traditionally, not a unique

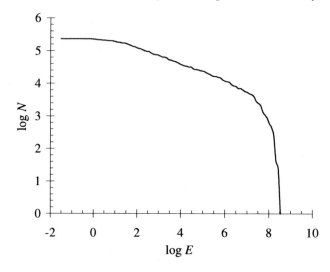

Figure 6
Cumulative frequency—energy diagram for the case of $\hat{l}_0 = 0.2$.

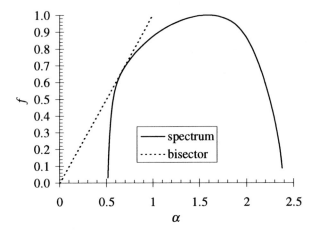

Figure 7
Multifractal spectrum of energy release ($\hat{l}_0 = 0.2$).

value, but an entire spectrum of β can be obtained from the seismic data treatment (KAGAN, 1991, 1994, 1997). Opinions exist that the scaling exponent(s) can be different not only for various seismic fields (FROHLICH and DAVIS, 1993; KAGAN, 1997), but even within the same set of seismic events. The uncertainty in determination of the unique scaling exponent presupposes the search for new methods for the characterisation of energy release (its temporal distribution). One such possibility is the use of the theory of multifractal spectra.

An additional method for a scaling analysis of the energy release with the crack propagation can be proposed with the utilisation of the multifractal formalism. Below a so-called canonical method is used (see, for instance, HALSEY et al., 1986; CHHABRA et al., 1989 and references therein) which provides a procedure for direct determination of the $f - \alpha$ relation, a singularity spectrum of q-th moments of distribution. This spectrum can be obtained as parametrically depending on q:

$$\alpha(q) = \lim_{\varepsilon \to 0} \frac{\sum_{i=1}^{\tilde{N}(\varepsilon)} \mu_i(q, \varepsilon) \log p_i(\varepsilon)}{\log \varepsilon}, \quad (11)$$

$$f(\alpha(q)) = \lim_{\varepsilon \to 0} \frac{\sum_{i=1}^{\tilde{N}(\varepsilon)} \mu_i(q, \varepsilon) \log \mu_i(q, \varepsilon)}{\log \varepsilon}. \quad (12)$$

For numerical calculations with the use of Eqs. (11), (12) the time interval from the first local failure (and of the respective energy release) until the end of the fault propagation is covered by $\tilde{N}(\varepsilon)$ segments ('boxes' in terms of the fractal theory) of length ε. Next, $p_i = E_i/E_{tot}$; $\mu_i = (p_i)^q / \sum_{i=1}^{N} (p_i)^q$; here E_i is the amount of released energy corresponding to the i-th segment ($i \in [1, \tilde{N}(\varepsilon)]$), E_{tot} is the total energy released under the entire process of fault propagation. Figure 7 presents results of the application of the canonical method [by means of excluding q from relations (11) and (12)] to data on energy release obtained by numerical simulation. The spectrum for released energy possesses all the properties of the multifractal one: it is a cup convex that lies under the bisector $f(\alpha) = \alpha$ and has only a single connection point to it; its maximum $\max_\alpha f(\alpha) = 1$ is equal to the dimension of the geometrical support of the set under study, the axis t. The approval of multifractal properties of the energy release in the crust during the growth of the pre-existing fault means that the moments of temporal distribution M_q of energy scale as a power law with an infinite set of exponents $\tau(q)$ (EVERTSZ and MANDELBROT, 1992):

$$M_q \propto \varepsilon^{-\tau(q)}. \qquad (13)$$

These exponents are connected with the $f(\alpha)$ function by the Legendre transform:

$$f(\alpha(q)) = q\alpha(q) - \tau(q). \qquad (14)$$

Conclusion

The proposed model of the shear crack propagation incorporates ideas of various methods used for the description of faults and seismicity. One of the principle moments is the concept of crack-damage interaction which allows introduction of the fracture processes at lower scale levels to be considered. The additional variable, damage parameter, describes the effect of these processes on the deformation evolution on the scale level under study. Unification of a continuum description with discretisation of the analysed region into elements serves as a basis for consideration of spatial randomness in crust properties. As a result, a tortuous shape of the fault front at propagation, which is considerably closer to real, complicated morphology of faults, is obtained in simulations. Both the width of the process zone and the amount of released energy demonstrate the scaling character. The former is characterised by the single scaling exponent, while the temporal distribution of the latter has multifractal properties.

Understanding that presentation of the seismic epicentre is such a simplified form is quite an inadequate description of the entire spectrum of phenomena linked with earthquakes, although it still allows the study of some of its features. Introduction of additional techniques for the characterisation of scaling features of energy release and its application to data obtained by means of numerical simulation, can serve as a step for the expansion of one more method for the estimation and analysis of seismicity.

REFERENCES

ALLÈGRE, C. J., LE MOUËL, J. L., CHAU, H. D., and NARTEAU, H. (1995), *Scaling Organization of Fracture Tectonics (SOFT) and Earthquake Mechanism*, Phys. Earth Planet. Inter. *92*, 215–233.

BURRIDGE, R., and KNOPOFF, L. (1967), *Model and Theoretical Seismicity*, Bull. Seismol. Soc. Am. *57*, 341–371.

CHEN, K., BHAGAVATULA, R., and JAYAPRAKASH, C. (1997), *Earthquakes in Quasistatic Models of Fracture in Elastic Media: Formalism and Numerical Techniques*, J. Phys. A: Math. Gen. *30*, 2297–2315.

CHHABRA, A. B., MANEVEAU, C., JENSEN, R. V., and SREENIVASAN, K. R. (1989), *Direct Determination of $f(\alpha)$ Singularity Spectrum and its Application to Fully Developed Turbulence*, Phys. Rev. A *40*, 5284–5293.

COWIE, P. A., and SCHOLZ, C. H. (1992), *Growth of Faults by Accumulation of Seismic Slip*, J. Geophys. Res. *97*, 11,085–11,095.

CURTIN, W. A., and SCHER, H. (1997), *Time-dependent Damage Evolution and Damage in Materials. I. Theory*, Phys. Rev. B *55*, 12,038–12,050.

CURTIN, W. A. (1998), *Size Scaling of Strength in Heterogeneous Materials*, Phys. Rev. Lett. *80*, 1445–1448.

EVERTSZ, C. J. G., and MANDELBROT, B. B., *Multifractal measures*. In *Chaos and Fractals. New Frontiers of Science* (eds. Peitigen, H. O., Jürgens, H., and Saupe, D.) (Springer, Berlin 1992) pp. 921–953.

FROHLICH, C., and DAVIS, S. D. (1993), *Teleseismic b Values; or, Much Ado about 1.0*, J. Geophys. Res. *98*, 631–644.

GROLEAU, D., BERGERSEN, B., and XU, H.-J. (1997), *Scaling Properties of a Model for Ruptures in an Elastic Medium*, J. Phys. A: Math. Gen. *30*, 3407–3419.

GUTENBERG, B., and RICHTER, F., *Seismicity of the Earth and Associated Phenomena* (Princeton University 1954).

HALSEY, T. C., JENSEN, M. H., KADANOFF, L. P., PROCACCIA, I., and SHRAIMAN, B. I. (1986), *Fractal Measures and Their Singularities: The Characterisation of Strange Sets*, Phys. Rev. A *33*, 1141–1151.

HERRMANN, H. J., and ROUX, S., *Statistical Models for the Fracture of Disordered Media* (North Holland, Amsterdam 1990).

KACHANOV, L. M., *Introduction to Continuum Damage Mechanics* (Nijhoff, Dordrecht, The Netherlands 1986).

KAGAN, Y. Y. (1991), *Seismic Moment Distribution*, Geophys. J. Int. *106*, 123–134.

KAGAN, Y. Y. (1994), *Observational Evidence for Earthquakes as a Nonlinear Dynamic Process*, Physica D *77*, 160–192.

KAGAN, Y. Y. (1997), *Seismic-moment—Frequency Relation for Shallow Earthquakes: Regional Comparison*, J. Geophys. Res. *102*, 2835–2852.

KRAJCINOVIC, D., *Damage Mechanics* (Elsevier, Amsterdam 1996).

LEMAITRE, J., and CHABOCHE, J.-L., *Mechanics of Solid Materials* (Cambridge University Press, Cambridge 1990).

LYAKHOVSKY, V., BEN-ZION, Y., and AGNON, A. (1997), *Distributed Damage, Faulting, and Friction*, J. Geophys. Res. *102*, 27,635–27,649.

LIEBOWITZ, H. (ed.), *Fracture*, Vol. I–VII (Academic Press, New York 1968–1972).

MURAKAMI, Y., *Stress Intensity Factors Handbook* (Pergamon Press, Oxford 1987).

SAHIMI, M., and ARBABI, S. (1996), *Scaling Laws for Fracture of Heterogeneous Materials and Rock*, Phys. Rev. Lett. *77*, 3689–3692.

SCHOLZ, C. H., *Earthquakes and faulting: self-organized critical phenomena with a characteristic dimension*. In *Spontaneous Formation of Space-time Structures and Criticality* (eds. Riste, T., and Sherrington, D.) (Kluwer Academic Press, Amsterdam 1991) pp. 41–56.

SCHOLZ, C. H., *Fractal transitions on geological surfaces*. In *Fractals in the Earth Sciences* (eds. Barton, C. C., and La Pointe, P. R.) (Plenum Press, New York 1995) pp. 131–140.

SCHOLZ, C. H. (1997), *Scaling Properties of Faults and Their Populations*, Int. J. Rock. Mech. and Min. Sci. *34*, paper No. 273.

SCHOLZ, C. H., DAWERS, N. H., YU, J.-Z., ANDERS, M. H., and COWIE, P. A. (1993), *Fault Growth and Fault Scaling Laws: Preliminary Results*, J. Geophys. Res. *98*, 21,951–21,961.

SILBERSCHMIDT, V. V. (1992), *The Fractal Characterization of Propagating Cracks*, Int. J. Fracture *56*, R33–R38.

SILBERSCHMIDT, V. V. (1993), *Analysis of rock fracture under the localization of shear deformations*. In *Assessment and Prevention of Failure Phenomena in Rock Engineering* (eds. Pasamehmetoglu, A. G., Kawamoto, T., Whittaker, B. N., and Aydan, Ö.) (A.A. Balkema, Rotterdam/Brookfield 1993) pp. 143–148.

SILBERSCHMIDT, V. V., *Fractal characteristics of joint development in stochastic rocks*. In *Fractals and Dynamic Systems in Geoscience* (ed. Kruhl, J. H.) (Springer-Verlag, Berlin 1994) pp. 65–76.

SILBERSCHMIDT, V. V. (1996), *Fractal Approach in Modelling of Earthquakes*, Geol. Rundsch. *85*, 116–123.

SILBERSCHMIDT, V. V. (1998), *Analysis of Shear Damage Accumulation*, Int. J. Damage. Mech. 7, in press.

SILBERSCHMIDT, V. V., and CHABOCHE, J.-L. (1994a), *Effect of Stochasticity on the Damage Accumulation in Solids*, Int. J. Damage Mech. *3*, 57–70.

SILBERSCHMIDT, V. V., and CHABOCHE, J.-L. (1994b), *The Effect of Material Stochasticity on Crack-damage Interaction and Crack Propagation*, Eng. Fract. Mech. *48*, 379–387.

SILBERSCHMIDT, V. V., and SILBERSCHMIDT, V. G. (1990), *Fractal Models in Rock Fracture Analysis*, Terra Nova *2*, 483–487.

XU, H.-J., BERGERSEN, B., and CHEN, K. (1992), *Self-organized Ruptures in an Elastic Medium: A Possible Model for Earthquakes*, J. Phys. A: Math. Gen. *25*, L1251–1258.

(Received March 3, 1998, revised June 13, 1998, accepted August 28, 1998)

 To access this journal online:
http://www.birkhauser.ch

Fractal Approach of Structuring by Fragmentation

CRISTIAN SUTEANU,[1] DOREL ZUGRAVESCU[1] and FLORIN MUNTEANU[1]

Abstract—The paper addresses the problem of fingerprints of fragmentation processes, showing that structuring by fragmentation can be detected and investigated fruitfully if approached with a proper methodology, of which fractal instruments represent an important part. Studied aspects include fragments size distributions, fragments size-position relations, and long-range correlations in fracture profiles and fracture patterns. The choice of experiments (comminution of flat samples and fragmentation by desiccation cracking) was directed by the aim to complement existing studies from the point of view of fragmentation energy application, being also intended to provide data from processes where different mechanisms are at work. Power-law fragments size distributions were found, but also fragments clusters on dominant size intervals that point towards a fractal character of the size distribution from the point of view of the positions of the distribution maxima. The self-affine character of fractures and of fracture patterns could be emphasised on certain scale intervals, separated by thresholds that are important for studies concerning the implied mechanisms. Fragments size-position correlations indicate a high probability of neighbouring fragments to be of comparable size. The features highlighted for structuring by fragmentation were found in all the studied experiments, their generality pointing towards useful implications for geoscientific research.

Key words: Rock fragmentation, crack patterns, fractal distributions, geodynamics.

1. Introduction

Since fragmentation phenomena proved to be significant for important geoscientific research fields (in geology, geophysics, geodynamics, planetology), considerable work was dedicated to their experimental investigation and modelling.

Among the studied aspects, fragments size distribution was widely investigated, due to its importance for the understanding of fragmentation processes in various types of context (FUJIWARA *et al.*, 1993; HOUSEN *et al.*, 1991; KATO *et al.*, 1995; NAKAMURA and FUJIWARA, 1991; TAKAGI *et al.*, 1984), to which one should add that the accessibility of this aspect also encouraged studies in this regard.

From the point of view of fragments size distribution, however, we witness a striking difference between the conclusions of two groups of works. On one hand, there is strong evidence for the fractal character of size distributions: the lack of a

---

[1] Romanian Academy, Institute of Geodynamics "Sabba S. Stefanescu," 19–21 Jean-Louis Calderon Str., Bucharest 37, 70201 Romania.

"characteristic size" is confirmed for a large variety of geomaterials and experimental conditions (TURCOTTE, 1986, 1993), size distributions having the form:

$$N(m) \propto m^{-w} \tag{1}$$

where $N(m)$ represents the number of fragments of mass larger than m.

On the other hand, extensive experimental works, performed also for a large diversity of fragmentation processes and of geomaterials, led other researchers (SADOVSKII et al., 1984, 1986) to the conclusion that every fragments size distribution is marked by a hierarchy of "characteristic sizes": these are emphasised by multimodal size distributions that point to a "hierarchic" character of "preferential dimensions" in the fragmented structure.

It should be noted here that both directions claim generality and that they deal, to a large extent, practically with the same materials, comminution conditions, etc. It was shown elsewhere (SUTEANU et al., 1993) that differences in the approach characterising the two directions, regarding mainly the data representation mode, can significantly contribute to the contradictions that arise between the two sets of conclusions. Special attention thus will be given here to the effects of data handling and representation methods.

We should emphasise that the question concerning the existence of dominant size intervals is meaningful and deserves special attention, especially in the case of complex systems in a critical state, like those studied in geodynamics (KING and SAMMIS, 1993; MUNTEANU et al., 1995; MAIN, 1996).

For this reason, we shall be interested here in the significance of fragmentation conditions (like fragmentation time, spatial energy concentration) from the point of view of their effect for a possible emergence of dominant size intervals.

Another aspect of solids fragmentation—less studied, despite its possibly high relevance for the behaviour of geosystems—concerns the relation between fragments sizes and their mutual spatial positions. Information concerning such a relation would be valuable in many situations that deal with fragmented structures, from geology and exploration geophysics to planetology (SUTEANU, 1996).

The fractal characterisation of fracture surfaces and crack patterns should highlight, on the other hand, features of fragmentation outcomes that would contribute to a better knowledge concerning these complex processes characterised by high variability.

These investigations will regard processes of rapid destruction, as well as those related to slow fissuring. The first case is relevant for studies concerning violent geodynamic events, like sudden rupture and comminution associated with earthquakes (SCHOLZ, 1990; MAIN, 1996), or for studies in planetology focusing on collisions of planetary bodies (HOUSEN and HOLSAPPLE, 1990; MILANI et al., 1994). The second case is interesting for many natural structures investigated within the framework of geology, exploration geophysics, etc. (BEN HASSINE and EL BORGI, 1994; SUTEANU et al., 1998a,b).

Given the fact that fragmentation processes imply a particular variability in their outcomes, a high sensitivity to a considerable number of factors (FINEBERG et al., 1991; FUJIWARA, 1991), we shall focus on the detection of certain fingerprints that would point, by their generality, towards useful implications for the studies regarding geosystems affected by fragmentation.

Our study will therefore approach, in this stage, the question of the structuring by fragmentation from the point of view of fragments sizes and their relation with their mutual spatial position, on one hand, and the description of fractures, on the other.

2. Fragmentation Experiments

2.1. Choice of Experiments

As stated in section 1, we are interested in the investigation of results of rapid destruction by shock, as well as in the study of structures obtained by slow fragmentation.

On the other hand, under the circumstances pointed out earlier—regarding the claimed general character of the contradicting conclusions about fragmentation outcomes—the investigation will have to approach experiments involving different types of rupture processes and, as far as possible, different types of mechanisms. We therefore shall study solids fragmentation by shock, but also a very different kind of fragmentation, occurring by desiccation cracking.

The chosen types of experimental context will also have to treat different situations from the point of view of the spatial concentration of fragmentation energy: since it is our aim to check whether fragmentation leads to general, recognisable properties, from the point of view of fragments size and position distributions, we shall treat experiments where fragmentation occurs in a different manner compared to the one studied in SADOVSKII (1986) and SADOVSKII et al. (1984). Fragmentation by collision between a target and a small projectile usually leads to the development of a highly differentiated fragmented structure from the point of view of the spatial distribution of parts: the region close to the contact area comprises mainly small fragments, while further from the impact location larger fragments may develop (FUJIWARA, 1991). It is therefore advisable to explore whether multimodal size distributions would arise even if fragmentation does not occur through point-like energy application. We therefore shall study cases in which the impact is applied on a large surface or where fragmentation energy is even more uniformly distributed.

For these reasons, we shall refer to two types of experiments:

A. Rapid fragmentation by shock applied to a large area (flat samples smashed against a plane surface).

B. Fragmentation by slow fissuring (desiccation of suspension films).

2.2. Experimental Methods

Experiments A—Rapid fragmentation by shock applied to a large area

Artificial flat samples (Romceram bricks) with a mass between 500 and 1200 g were smashed against a plane surface, the impact area representing 35–40% of the total sample surface. Fragmentation energy was in each case 60 J/kg. The impact was realised by dropping the samples horizontally on a flat slab. To insure the horizontal position of the sample in the moment of the impact, an orientation device was used (SUTEANU, 1997): a thin polyethan bag contained the sample. The upper part of the bag was dragged due to air friction, during the fall, towards an upright position, while the sample on its base preserved its horizontality (Fig. 1). Dropping occurred only after the initial horizontal equilibrium position was reached.

Fragmentation outcomes were evaluated by weighing the fragments on an electronic device (precision 0.00001 g) and by subsequent body reconstruction (section 4).

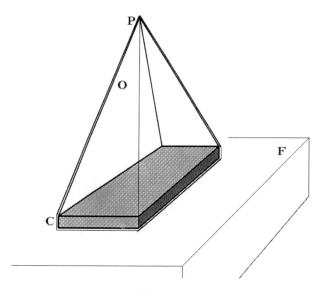

Figure 1
Experimental setup for fragmentation by shock applied to a large area. The flat sample C was laid on the bottom of the polyethan bag O and suspended by its apex P. The polyethan bag served as an orientation device, being dragged by air friction upwards and maintaining the sample C in a horizontal position. The bag with the sample was lifted by point P and, after the initial horizontal equilibrium position was reached, it was dropped on the concrete plate F.

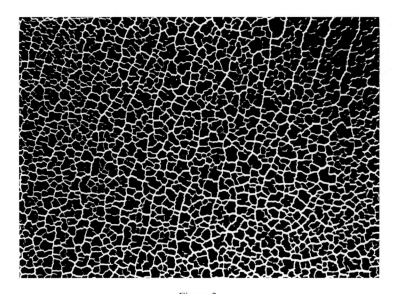

Figure 2
Fragmented structure obtained by desiccation: 50% starch suspension in water placed on a glass plate without lateral constraints (long edge size of the image is 50 mm).

Experiments B—Fragmentation by slow fissuring

As argued in section 1, desiccation cracking was chosen for this research mainly due to two reasons: the fact that very different phenomena are at work (compared to solids fragmentation by shock) and the considerably higher uniformity of the energy application than the one obtained in other experiments. Desiccation cracking has been studied especially from the point of view of the generated fracture patterns (ALLAIN and PARISSE 1996; SKJELTORP and MEAKIN, 1988; SUTEANU et al., 1995). For the aims of this study, we are interested not only in the fractal aspects of the crack pattern, but also in the size distribution of independent fragments and in their size-position relation.

To this end, desiccation cracking of suspension films was first tested on a wide diversity of materials (clay, starch, grinded coffee) (SUTEANU et al., 1995; SUTEANU, 1997). For the purpose of this research we chose to use starch suspension (50%) in water, due to the good reproducibility and to the fact that it leads to clear structures suitable also for an evaluation from the point of view of independent fragments (Fig. 2). The resulting viscous liquid (starch suspension) was placed on transparent glass plates and let to spread without lateral constraints. Desiccation temperature was 23 C.

The fragmented structures were then evaluated on digitised photographs.

3. Fragments Sizes

3.1. Experiments A

The outcomes of the comminution process were evaluated by weighing (sect. 2). Cumulative log-log representations of the fragments size distributions led to the distinction of two scaling intervals (Fig. 3). For each interval, we found a distribution law characterised by eq. (1). We denoted by D_1 and D_2 the values

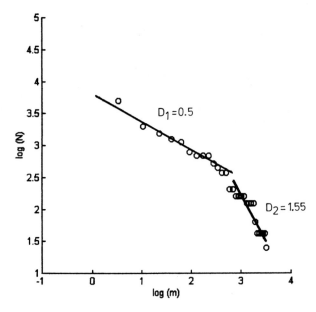

Figure 3
Fragments mass distribution obtained in experiments involving fragmentation by shock applied to a large area. Cumulative log-log representation (m = fragments mass, N = cumulative fragments number; D_1 and D_2 are obtained by linear interpolation for the two regimes).

Table 1

Results for fragmentation experiments of type A. (m = target mass in grams; D_1, D_2 = exponents of the fragments size distribution for the two regimes; R = mass recovery rate)

Parameters Sample no.	$m[g]$	D_1	D_2	R [%]
1	532	0.5	1.55	95
2	680	0.45	1.52	98
3	982	0.42	1.55	94
4	548	0.38	1.6	97
5	572	0.4	1.56	96
6	1230	0.3	1.45	90
7	1085	0.5	1.30	92

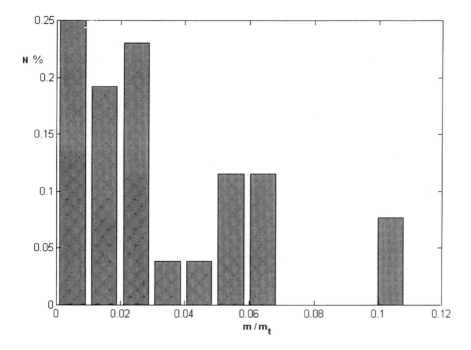

Figure 4
Fragments size distribution concerning fragmentation by shock applied to a large area. Linear, non-cumulative representation that emphasizes the multimodal character of the size distribution (m = fragments mass, m_t = total mass of the initial body; N = percentage of the total number of fragments).

corresponding to the exponent w in eq. (1) for each of the distinct intervals. Table 1 offers a synthetic view over the results obtained.

Another aspect noticeable in these representations concerns the presence of "steps," local departures from the smooth power law, reminding those reported by KRUHL and NEGA (1996) for grain boundary suturing. Since cumulative log-log representations strongly attenuate such departures from the power-law behaviour, it may also be useful to analyse the size distributions in linear scale (SUTEANU, 1996).

A linear non-cumulative representation of the fragments size distribution reveals a clustering of fragments on certain size intervals. This clustering (Fig. 4) was found in all studied cases.

As stated in section 1, this aspect had already been announced by a group of researchers (SADOVSKII, 1986; SADOVSKII et al., 1984) who interpreted the multi-modal distribution as an expression of a series of "preferential dimensions" that obey a geometric progression.

Our studies point towards the fact that both aspects—the overall fractal character of the distribution, as shown by cumulative log-log plots, and its multimodal character—can be simultaneously valid. Moreover, the way the distribution modi emerge, with a tendency to correspond to a power law, suggests that size distributions may feature one more fractal aspect: that is a fractal character of the distribution structure, as revealed in linear non-cumulative plots. Distributions generated for the same fragments set with growing resolution emphasise the emergence of maxima, arranged in a way that approximates, indeed, a succession corresponding to a power law (Figs. 5a, b).

3.2. Experiments B

Cumulative log-log plots of the fragments size distributions also show size intervals that correspond to a power-law distribution (Fig. 6):

$$N(A) \propto A^{-q} \qquad (2)$$

where N is the number of fragments with a size greater than A.

The values for the exponent q obtained in our experiments (performed under the circumstances described in section 2.2) are given in Table 2.

Linear plots of the size distributions emphasise dominant size intervals too. Despite the more uniform fragmentation energy application (ALLAIN and PARISSE, 1996) and the different fragmentation mechanisms that act in the case of desiccation cracking (SKJELTORP and MEAKIN, 1988), we also encounter here size intervals corresponding to fragments clustering. Figure 7 shows the graphs of three distinct areas obtained in the same experiment. One may notice that, despite visible differences in the details of the distributions, there are common intervals corresponding to fragments clusters.

4. Fragments Size-position Correlation

As pointed out in section 1, knowledge of a relation between fragments' sizes and their mutual spatial positions would significantly enhance the practical importance of the research concerning structuring by fragmentation. We studied this aspect in both types of fragmentation experiments.

4.1. Experiments A

Three-dimensional body reconstruction was used for the purpose of this research. We computed the ratio p between masses of adjacent fragments. The distribution of the number of ratios $R(p)$ larger than p demonstrated that the probability of two adjacent fragments to be of very different size (large p) rapidly decreases with growing p. In the case of fragmentation by shock, this decrease obeys an exponential law:

$$R(p) \propto e^{-\alpha p} \qquad (3)$$

with $\alpha \in [1, 2]$ (Fig. 8, $\alpha = 1.6$).

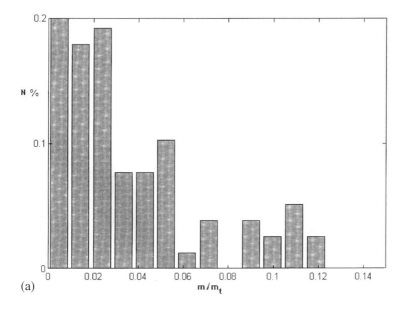

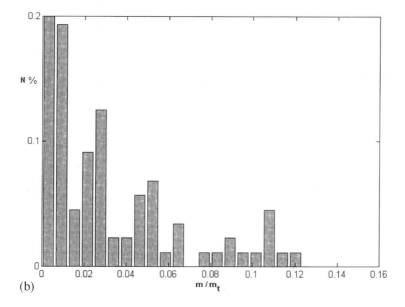

Figure 5
a, b. Fragments size distributions corresponding to one and the same fragments set obtained in experiments involving shock applied to a large area. The way maxima emerge with growing resolution suggests a tendency towards a self-similar structure of the distribution (m = fragments mass, m_t = total mass of the initial body; N = percentage of the total number of fragments).

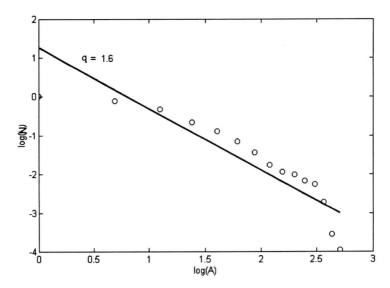

Figure 6
Fragments size (area) distribution corresponding to an experiment of fragmentation by desiccation. Cumulative, log-log representation (A = fragments area; N = percentage of the total number of fragments; q was determined by linear interpolation).

Table 2

Results for fragmentation experiments of type B. (q = exponent of the fragments size distribution)

Sample no.	1	2	3	4	5	6	7	8	9	10
q	1.91	2.00	1.95	1.65	1.81	1.84	1.78	1.75	1.82	1.88

4.2. Experiments B

In the case of desiccation cracking, the tendency of fragments of comparable size to have neighbouring positions is also present, but the decrease of the number of ratios $R(p)$ obeys a power law:

$$R(p) \propto p^{-\eta} \qquad (4)$$

with $\eta \in [1, 2]$ (Fig. 9, $\eta = 2$).

While in the two situations (experiments A and B) we witness a distinct behaviour from the point of view of the type of decrease of the function $R(p)$, they share an important point: in both cases, the small ratios are more frequent than the large ones, showing a tendency of neighbouring fragments to be of comparable sizes.

5. Fractures

The high irregularity of crack surfaces and fracture patterns can be fruitfully approached with the help of fractal methodology (KAYE, 1986, 1989; KRUHL, 1994; TAKAYASU, 1992; GOUYET, 1992; TURCOTTE, 1993). From the point of view of the studies concerning fragmented structures, the investigation of fractures represents "the other side of the coin" than the one that focuses on fragments sizes. Given its significance for the systems undergoing rupture or shaped by fragmentation processes, this aspect deserves special attention.

The two sets of experiments will be treated according to a specific methodology.

5.1. Experiments A

We analysed rupture profiles of the fragments produced by comminution. To this end, we used digitised photographs of the intersection between the fracture surface and a perpendicular plane. Figure 10 shows one of the profiles.

A reliable method to determine the fractal dimension of such a profile is the "height-height correlation" method, which relies on the correlation moment of second order (FEDER, 1988; BOUCHAUD et al., 1993; BOUCHAUD and BOUCHAUD, 1994). One starts from the correlation moment of second order:

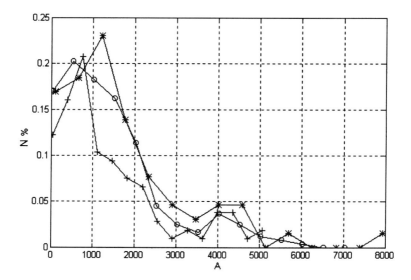

Figure 7

Fragments size distributions corresponding to three distinct zones of the same structure produced in a desiccation experiment (linear, non-cumulative representation). Lines between the points are only meant to guide the eye. Despite differences in the distribution shapes, all of them highlight clusters on the same size intervals. (A = fragments area in units equal to 0.01 mm^2; N = percentage of the total number of fragments).

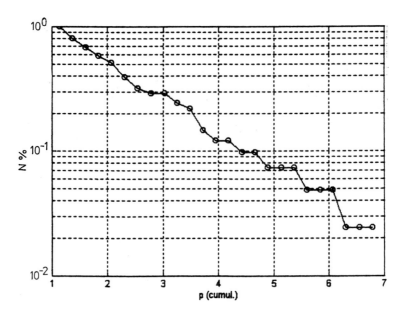

Figure 8
Distribution corresponding to the ratios p between masses of adjacent fragments obtained in experimental fragmentation by shock applied to a large area. An exponential decrease of the cumulative number of ratios p is visible (N = percentage of the total number of ratios).

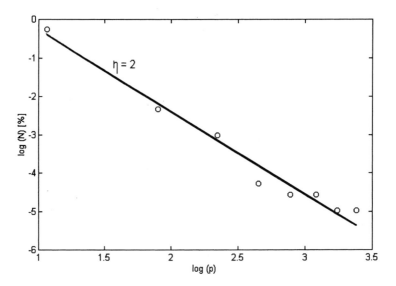

Figure 9
Distribution corresponding to the ratios p between sizes of adjacent fragments obtained in experimental fragmentation by desiccation. A power-law decrease with an exponent equal to 2 is emphasised.

$$M_2(r) = \langle [z(x+r) - z(x)]^2 \rangle_x \tag{5}$$

where x is the value of the abscissa (along the profile), $z(x)$ is the profile height and r is a variable distance separating the pair of points considered for the computing of the correlation. The symbols $\langle$ and $\rangle$ show that the average is computed for all values of x.

In a next step, the existence of the following relation is checked:

$$M_2(r) \propto r^{2\lambda} \tag{6}$$

and, if fulfilled, this relation leads to the fractal dimension of the profile:

$$\delta = 2 - \lambda \tag{7}$$

from which one obtains the fractal dimension of the fracture surface (MANDELBROT, 1982):

$$D = \delta + 1. \tag{8}$$

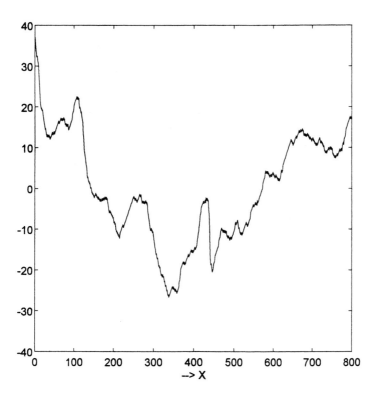

Figure 10
Rupture profile obtained in experiments of fragmentation by shock (vertical exaggeration × 10). Abscissa: profile length in units of 0.1 mm.

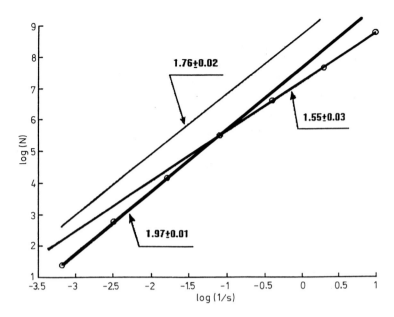

Figure 11
Results of the fractal analysis (box-counting method) of a fragmentation pattern obtained in the desiccation experiments. Two scaling regimes are detectable, separated by a threshold corresponding to a box size s of 3 mm.

Applied to the fracture profiles obtained in our experiments, the method leads to the conclusion that the fractal character is detectable on a scale range with a lower limit of 3 mm. The upper limit is larger than the investigated limit of 30 mm (imposed by the fragments size). The resulting fractal dimension is: $D = 2.22 \pm 0.04$.

5.2. Experiments B

We were interested, in this case, in an evaluation of the complex crack pattern resulting from a fragmentation process like the one illustrated in Figure 2. For this type of investigation (two-dimensional, complex structure), one of the relevant and reliable methods is the box-counting method (TAKAYASU, 1992). In principle, the method relies on the determination of the way the number of "cells" $N(s)$ that cover the given structure scales with the side s of the cells:

$$N(s) \propto s^{-D}. \qquad (9)$$

Applied to digitised photographs of fragmented structures, the method revealed in each case a fractal character (Fig. 11), in agreement with results obtained in similar experiments by other researchers (SKJELTORP and MEAKIN, 1988; GOUYET, 1992).

Nevertheless, a careful analysis based on a best-fit evaluation revealed the presence of two distinct regimes. While the overall fractal dimension, for the whole scale interval, is $D = 1.76 \pm 0.02$, we find that a threshold of $s = 3$ mm separates two regions with distinct spatial correlations: one with $D = 1.97 \pm 0.01$ (for scales larger than the threshold) and one with $D = 1.55 \pm 0.03$ (for smaller scales). Both of these regimes are characterised by a very good correlation of the power-law dependence in eq. (9) (a correlation coefficient better than 99.99%).

In our view, the "average" value $D = 1.76$ is in this case less relevant than the existence of the two regimes: while the first of them points towards the tendency of "plane filling" of the structure above a certain scale, the second regime offers information about the shape and spatial arrangement of the fracture lines on a scale interval that could be interesting for the characterisation of the fracturing process (SUTEANU et al., 1995). The limit that separates the two scaling regimes can also be important for the characterisation of a fragmented structure.

6. Discussion

Our experiments demonstrate that the presence of distinct multiple modi, found in all the studied fragments size distributions, does not necessarily collide with the assertion concerning the fractal character of these distributions. Both aspects, overall power-law dependence of the cumulative distribution envelope, on one hand, and a series of (correlated) dominant size intervals in linear non-cumulative plots of the distribution, on the other, can be emphasised for the same experiments. Data representation largely influences the aspect highlighted best.

Moreover, our studies suggest that the fractal character of the size distribution may extend beyond the aspect linked with the general shape of the cumulative distribution, governing also the succession of distribution maxima (the position of dominant size intervals). A claim concerning the general character of this new aspect of fractal behaviour would have to rely on more investigations, on a large variety of materials and process parameters, that would yield a rich population of fragments, enabling a relevant and detailed analysis, with variable resolution, of the distribution structure. Nevertheless, present studies strongly point towards this possibility, being also at this point in agreement with the experimental evidence of SADOVSKII (1986) and SADOVSKII et al. (1984), which does not mean that our studies support the presence of "preferential" dimensions. We did not find preferential absolute dimensions, the dominant size intervals depending upon the fragmented body mass.

On the other hand, we showed that in all the studied cases adjacent fragments express the tendency to have comparable sizes. These results are surprising if compared to largely accepted statements relying on geometric models like the Sammis comminution model (TURCOTTE, 1993): starting from the premise that it is not likely that a small fragment (block) can break a large neighbour, or that a large fragment can break a small neighbour, Sammis suggests that neighbouring fragments should not be of comparable size. Our experimental results point out that this assertion is not confirmed in practice: neighbouring blocks (fragments) tend to be of comparable size. We consider that we should not discuss the fragmentation process in terms of one fragment breaking its neighbouring fragment: the whole system responds to the applied stress by turning into a structure, a set of fragments, with certain properties, partly emphasised above. Why this complex process leads to this type of correlations is another question that cannot be discussed here in detail, we shall only point towards certain attempts to model fragmentation processes according to different types of approaches (BENZ and ASPHAUG, 1994; SUTEANU, 1997). On the other hand, Sammis' model was considered a success also because it led to an exponent in good agreement with observations in the field. One may argue of course that there are different spatial distributions of fragments, corresponding to different models, which would lead to the same distribution exponent. Therefore, the Sammis comminution model is not necessarily supported by field data *in its details*. A larger diversity of comminution processes, including those occurring in the field, should be analysed, before reaching a firm conclusion in this regard. At this stage, we must say, however, that our experiments (of both categories) point to a different conclusion than Sammis' model does.

The presence of distinct scaling regimes in the case of crack patterns also points towards the possibility of obtaining by fractal analysis useful information concerning complex structures: a careful approach, considering the possible existence of different regimes, may prove to be fruitful. We highlight here that the threshold between the scaling regimes could be as important as the scaling exponents and even more relevant for the studies concerning the scale ranges on which different mechanisms dominate the fracturing process.

7. Conclusions

Our studies highlight certain properties of fragmented structures which support the idea that fragmentation processes produce, despite their well-known variability, structures marked by certain general fingerprints. Power-law fragments size distributions, fragments clusters on dominant size intervals, self-affine character of fractures and fracture patterns on certain scale intervals and fragments size-position correlations that indicate a higher probability of neighbouring fragments to be of comparable size, are features confirmed by all our experiments.

Structuring by fragmentation can thus be detected and investigated fruitfully if approached with a proper methodology. Applied with rigour—always discerning the proper limits and taking care of aspects possibly obscured by the methods applied—fractal instruments prove capable of bringing important benefits.

The properties discussed could thus be fruitfully studied in different areas of the geosciences. Both the fragments clustering on certain size intervals and the spatial agglomeration of fragments of comparable size, are important for geodynamics, where, if added to other scaling properties studied to date (fault patterns, hypocenter distributions) they may considerably support geodynamic system modelling (SUTEANU, et al., 1997). Their implications are now investigated by the authors within the framework of studies pertaining to the seismogenic region of Vrancea, Romania.

Acknowledgements

The authors would like to thank Jörn H. Kruhl for fruitful discussions and enlightening comments, and two anonymous reviewers for their useful observations.

This work was performed within the framework of the project IG01.3 of the Institute of Geodynamics "Sabba S. Stefănescu" of the Romanian Academy.

REFERENCES

ALLAIN, C., and PARISSE, F. (1996), *Des fractures bien rangees*, La Recherche *288*, 50.
BEN HASSINE, K., and EL BORGI, M. (eds.) (1994), *Fractured Reservoir*, Tunis, ETAP.
BENZ, W., and ASPHAUG, E. (1994), *Impact Simulations with Fracture: I. Method and Tests*, Icarus *107*, 98–116.
BOUCHAUD, E., LAPASSET, G., PLANÈS, J., and NAVEOS, S. (1993), *Statistics of Branched Fracture Surfaces*, Phys. Rev. *B48* (5), 2917–2928.
BOUCHAUD, E., and BOUCHAUD, J.-P. (1994), *Fracture Surfaces: Apparent Roughness, Relevant Length Scales, and Fracture Toughness*, Phys. Rev. *B50* (23), 17,752–17,755.
FEDER, J., *Fractals* (New York, Plenum 1988).
FINEBERG, J., GROSS, S. P., MARDER, M., and SWINNEY, H. (1991), *Instability in Dynamic Fractures*, Phys. Rev. Lett. *67* (4), 457–460.
FUJIWARA, A., *Catastrophic disruption of solid bodies by collision—experimental approach*. In *Origin and Evolution of Interplanetary Dust* (eds. Levasseur-Regourd, A. C., and Hasegawa, H.) (Kluwer, Dordrecht 1991) pp. 361–366.

FUJIWARA, A., NAKAMURA, A., KATO, M., and TAKAGI, Y., *Experimental simulation of collisions*. In *Primitive Solar Nebulae and Origin of Planets* (ed. Oya, H.) (Terra Scientific, Tokyo 1993) pp. 281–295.

GOUYET, J. F., *Physique et structures fractales* (Paris, Masson 1992).

HOUSEN, K. R., and HOLSAPPLE, K. A. (1990), *On the Fragmentation of Asteroids and Planetary Satellites*, Icarus *84*, 226–253.

HOUSEN, K. R., SCHMIDT, R. M., and HOLSAPPLE, K. A. (1991), *Laboratory Simulation of Large-scale Fragmentation Events*, Icarus *94*, 180–190.

KATO, M., IIJIMA, Y. I., ARAKAWA, M., OKIMURA, Y., FUJIMURA, A., MAENO, N., and MIZUTANI, H. (1995), *Ice-on-ice Impact Experiments*, Icarus *113*, 423–441.

KAYE, B. H., *Fractal geometry and the characterization of rock fragments*. In *Fragmentation, Form and Flow in Fractured Media* (eds. Englman, R., and Jaeger, Z.) (Adam Hilger, Bristol 1986) pp. 490–516.

KAYE, B. H., *A Random Walk through Fractal Dimensions* (Weinheim, VCH 1989).

KING, G. C. P., and SAMMIS, C. G., *The mechanisms of finite brittle strain*. In *Fractals and Chaos in the Earth Sciences* (eds. Sammis, C. G., Saito, M., and King, G. C. P.) (Birkhäuser, Berlin 1993) pp. 611–640.

KRUHL, J. H. (ed.) *Fractals and Dynamic Systems in Geoscience* (New York, Springer 1994).

KRUHL, J. H., and NEGA, M. (1996), *The Fractal Shape of Sutured Quartz Grain Boundaries: Applications as a Geothermometer*, Geol. Rundsch. *85*, 38–43.

MAIN, I. G. (1996), *Statistical Physics, Seismogenesis, and Seismic Hazard*, Reviews of Geophysics *34* (4), 433–462.

MANDELBROT, B. B., *The Fractal Geometry of Nature* (San Francisco, John Wiley 1982).

MILANI, A., DI MARTINO, M., and CELLINO, A. (eds.) *Asteroids, Comets, Meteors* (Dordrecht, Kluwer 1994).

MUNTEANU, F., IOANA, C., SUTEANU, C., and ZUGRAVESCU, D. (1995), *Discriminating Transient Dynamics and Critical States in Active Geodynamic Areas*, Studii si Cercetari de Geofizica *33*, 10.

NAKAMURA, A., and FUJIWARA, A. (1991), *Velocity Distribution of Fragments Formed in a Simulated Collisional Disruption*, Icarus *92*, 132–146.

SADOVSKII, M. A., GOLUBEVA, T. V., PISARENKO, V. F., and SHNIRMAN, M. G. (1984), *Characteristic Dimensions of Rock and Hierarchical Properties of Seismicity*, Izv. Acad. Sci. USSR Phys. Solid Earth, Engl. transl. *20*, 87–96.

SADOVSKII, M. A. (1986), *Some Results in Seismology Obtained on the Basis of the New Media Model*, Bulg. Geophys. J. *3* (12), 3–7.

SCHOLZ, C. H., *The Mechanics of Earthquakes and Faulting* (New York, Cambridge University Press 1990).

SKJELTORP, A., and MEAKIN, P. (1988), *Fracture in Microsphere Monolayers Studied by Experiment and Computer Simulation*, Nature *335*, 6189, 424–426.

SUTEANU, C. (1996), *Structuring by Fragmentation Revealed by 3-dimensional Evaluation*, Studii si Cercetari de GEOFIZICA *34*, 2–14.

SUTEANU, C. (1997), *Hierarchies and Invariance in Solids Fragmentation. Implications for Geodynamics*, Ph.D. Thesis, Romanian Academy, Institute of Geodynamics, Bull. Inst. Geodyn. *9*(3), 1–343.

SUTEANU, C., IOANA, C., MUNTEANU, F., and ZUGRAVESCU, D. (1993), *Fractal Aspects in Solids Fragmentation. Experiments and Model with Implications for Geodynamics*, Révue Roumaine de Géophysique *37*, 61–79.

SUTEANU, C., MUNTEANU, F., and ZUGRAVESCU, D. (1995), *Hierarchies, Scaling and Anisotropy in Dehydration Cracking*, Révue Roumaine de Géophysique *39*, 3–11.

SUTEANU, C., MUNTEANU, F., and ZUGRAVESCU, D. (1997), *Scaling Regimes and Anisotropy: Towards an Effective Approach to Complex Geologic Structures*, Révue Roumaine de Géophysique *41*, 25–43.

SUTEANU, C., ZUGRAVESCU, D., IOANA, C., and MUNTEANU, F. (1998a), *Fracture morphology and fluid flow properties in fractured reservoir: Tools for strutural characterization and prediction*, Proceedings of the African/Middle East Second International Geophysical Conference, Cairo, Egypt, 17–19 Febr. 1998.

SUTEANU, C., IOANA, C., MUNTEANU, F., and ZUGRAVESCU, D. (1998b), *Scaling Aspects in the Collisional History of Planetary Bodies*, Révue Roumaine de Géophysique *42*, 15–26.

TAKAGI, Y., MIZUTANI, H., and KAWAKAMI, S. (1984), *Impact Fragmentation Experiments of Basalts and Pyrophyllites*, Icarus *59*, 462–467.

TAKAYASU, H., *Fractals in the Physical Sciences* (Wiley, New York 1992).

TURCOTTE, D. L. (1986), *Fractals and Fragmentation*, J. Geophys. Res. *91*, 1921–1926.

TURCOTTE, D. L., *Fractals and Chaos in Geology and Geophysics* (Cambridge University Press, Cambridge 1993).

(Received May 21, 1998, accepted July 14, 1999)

To access this journal online:
http://www.birkhauser.ch

Micromorphic Continuum and Fractal Fracturing in the Lithosphere

H. NAGAHAMA[1] and R. TEISSEYRE[2]

Abstract—It seems that internal structures and discontinuities in the lithosphere essentially influence the lithospheric deformation such as faulting or earthquakes. The micromorphic continuum provides a good framework to study the continuum with microstructure, such as earthquake structures. Here we briefly introduce the relation between the theory of micromorphic continuum and the rotational effects related to the internal microstructure in epicenter zones. Thereafter the equilibrium equation, in terms of the displacements (the Navier equation) in the medium with microstructure, is derived from the theory of the micromorphic continuum. This equation is the generalization of the Laplace equation in terms of displacements and can lead to Laplace equations such as the local diffusion-like conservation equations for strains. These local balance/stationary state of strains under the steady non-equilibrium strain flux through the plate boundaries bear the scale-invariant properties of fracturing in the lithospheric plate with microstructure.

Key words: Micromorphic continuum, rotational wave, faults, earthquakes, fractals.

Introduction

The internal structures and discontinuities in the lithosphere seem to essentially influence its fracturing. Thus, it is reasonable to believe that the notion of continuum with microstructure can be a suitable tool in describing earthquake phenomena. The scale of earth structures with their extremely complicated internal microstructures justifies the use of continuum mechanics (TEISSEYRE, 1973a,b, 1982, 1995a,b). Particularly the generalized micromorphic continuum is suitable to introduce microstructure (SUHUBI and ERINGEN, 1964; ERINGEN and SUHUBI, 1964; ERINGEN, 1968; ERINGEN and CLAUS, 1970). When the deformations imposed on a microstructural element favour its elongation rather than rotation during fracturing, the symmetric micromorphic continuum is suitable to our considerations of the role of microstructure in the region of the earth interior where earthquakes occur (TEISSEYRE, 1973a,b, 1974, 1975, 1978, 1982, 1986, 1995a,b).

---

[1] Institute of Geology and Paleontology, Graduate School of Sciences, Tohoku University, Sendai 980-8578, Japan.
[2] Institute of Geophysics, Polish Academy of Sciences, ul. Księcia Janusza 64, 01-452 Warzaw, Poland.

However, when the friction motion along the precuts occurs during the earthquakes, it is better to use the asymmetric stress micromorphic continuum (similar to the micropolar mechanics: SHIMBO, 1978; IESAN, 1981; TEISSEYRE, 1995a,b).

Several scaling properties of faults or earthquakes, such as Gutenberg-Richter's relation, have been studied (e.g., TURCOTTE, 1986a,b,c, 1992; NAGAHAMA and YOSHII, 1994). By considering the model of self-organized criticality (SOC model; BAK et al., 1987, 1988; ZHANG, 1989; HWA and KARDAR, 1989), it was found that if the dynamics satisfies a local conservation law, then the steady configurations are to be fractal or the system will be self-organized into a critical state. In another SOC model of earthquakes, ITO and MATSUZAKI (1990) applied ENYA'S (1901) idea that the main shock disturbs and the aftershocks occur to decrease the heterogeneity of strain distrubution in the crust, and derived some scale-invariant properties of earthquakes. Moreover, for an SOC model for the long-term deformations of the lithospheric plate, D. Sornette and coworkers also presented a new diffusion-like strain-governing equation which hypothesizes that the steady flow of tectonic stresses (strains) generates the fractal nature of earthquakes (SORNETTE D. et al., 1990; SORNETTE and VIREUX, 1992; SORNETTE and SORNETTE, 1994). However, this diffusion-like strain-conservation equation was not based on any concrete theory of the micromorphic continuum or rheology of the lithospheric plate with microstructure.

The average displacement within the plate caused by the tectonic stress flow through the plate's boundaries has been studied by researchers (ELSASSER, 1969; RICE, 1980; LEHNER et al., 1981; LI and RICE, 1983). These studies should be focused on the macroscopic deformation of the lithospheric plate (non-equilibrium macroscopic flux of the strains). On the other hand, the approaches based on the micromorphic continuum for the earthquakes have presented the equilibrium equation in terms of displacement within the lithospheric materials with microstructures (the local balance/stationary state of strains). Here, we can encounter a paradox between the two states of strains. This paradox has not been thoroughly discussed yet from the view-point of continuum mechanics. Relations between the theory of micromorphic continuum and the fractal properties of fracturing in the lithosphere, such as faults or earthquakes, have been inadequately studied.

In this paper, the theory of the micromorphic continuum and rotational effects related to the internal microstructures at the epicenter zones are briefly introduced and the equilibrium equations in terms of displacements (the Navier equation) are derived from the concept of this micromorphic continuum. Thereafter we point out that the material constants of micromorphic continuum changes with the deformation like self-organization. Next it is shown that these equations are the generalization of Laplace equations in terms of displacements which generate several local diffusional conservation equations (Laplace equations) for strains within the lithospheric plate with microstructures. Moreover, a generalization of Elsasser and Rice's model of stress-diffusion (ELSASSER, 1969; RICE, 1980) and some scaling

Micromorphic Continuum

In this section we will briefly introduce the micromorphic continuum theory developed by Eringen and coworkers (SUHUBI and ERINGEN, 1964; ERINGEN and SUHUBI, 1964; ERINGEN, 1968; ERINGEN and CLAUS, 1970). In further calculations we will confine ourselves only to the linear theory and Cartesian coordinate system.

In the micromorphic continuum theory the deformations are represented not only by the displacement vector **u**, but also by a new tensor which describes deformations and rotations of microelements (grains, blocks or interstitial surface defects). It is a microdisplacement tensor φ. The deformation now can be expressed by (e.g., ERINGEN, 1968):

$$\text{strain tensor} \quad e_{ik} = u_{(i,k)}, \tag{1}$$

$$\text{microstrain tensor} \quad \varepsilon_{nl} = u_{l,n} + \varphi_{nl}, \tag{2}$$

$$\text{microdislocation density} \quad \Lambda_{kl} = -\epsilon_{lmn}\varphi_{kn,m} \tag{3}$$

where ϵ_{lmn} is Eddington's epsilon (the skew-symmetric tensor; 0, 1, -1). Note that we used here the following abbreviated notation: $u_{i,k} \equiv \partial u_i/\partial x_k$;

$$u_{(i,k)} \equiv \frac{1}{2}(u_{i,k} + u_{k,i}).$$

Taking into account the value of φ, we can distinguish three cases (e.g., BIELSKI, 1995):

(i) φ has nine independent components, and then each particle has locally 12 degrees of freedom,
(ii) $\varphi = \varphi^T$ is a symmetric tensor with six independent components and each particle has 9 degrees of freedom, and finally,
(iii) $\varphi = -\varphi^T$, i.e., φ is an antisymmetric tensor which can be represented by three independent components.

In the last case the micromorphic medium is reduced to a micropolar one (TEISSEYRE, 1973a; TEISSEYRE and NAGAHAMA, 1999).

Moreover, the elements forming the micromorphic continuum are characterized by their microinertia properties (TEISSEYRE, 1973a; TEISSEYRE and NAGAHAMA, 1999). These are described by the microinertia tensor

$$N_{ik} = \frac{1}{\rho v} \iiint_v \xi_i \xi_k (\rho_1 dv_1 + \rho_2 dv_2) \quad (4)$$

where the two medium constitutes have densities ρ_1, ρ_2, occupying volume v_1 and v_2 ($v = v_1 + v_2$), ξ_i are coordinates of body elements from its mass center, ρ is average density. N_{kl} is related to standard rigid body inertia J_{kl} by the relation

$$J_{kl} = N_{nn}\delta_{kl} - N_{kl} \quad (5)$$

where δ_{kl} is Kronecker's delta (0 if $k \neq l$ and 1 if $k = l$). We limit our considerations only to diagonal terms of microinertia

$$N_{kl} = N_{\langle kk \rangle}\delta_{kl} \quad (6)$$

where the symbol $\langle kk \rangle$ means that summation convention for indexes kk is not applied. From the balance equation of the considered continuum fields, the relations between the microinertia tensor and the microdisplacements (TEISSEYRE, 1973a):

$$N_{\langle kk \rangle}\varphi_{lk} = N_{\langle ll \rangle}\varphi_{kl} \quad \text{or} \quad \varphi_{lk} = \frac{N_{\langle ll \rangle}}{N_{\langle kk \rangle}}\varphi_{kl}. \quad (7)$$

To complete our model of earthquake structure, we demand that deformations imposed on a microstructural element favour its elongation rather than rotation. This justifies our choice of micromorphic continuum in which length deformations of the Cosserat directors are allowed. In particular, when considering a focal region and its deformations, the micropolar theory (rotation only: $\varphi_{ns} = -\varphi_{sn}$) is rather inadequate for seismological problems (TEISSEYRE, 1973a,b, 1974, 1975, 1978, 1982, 1986, 1995a,b; TEISSEYRE and NAGAHAMA, 1999). However when considering the friction motion along the precuts, the asymmetric micromorphic continuum is more suitable (similar to the micropolar mechanics: SHIMBO, 1978; IESAN, 1981; TEISSEYRE, 1995a,b). Next we will focus on the rotational effects in the epicenter zones.

Rotational Effects at the Epicenter Zones

The internal microstructures in the epicenter zone essentially influence the seismic motions. Equation (7) means that even if stresses give zero rotation, the deformations can form microrotations and rotational waves can propagate. This explains how the microrotations can be included in the state of deformation, especially in the near source zone. The microrotations appear due to differences in inertia properties of the microelements permeating the continuum. Such a contribution to the rotation of elements is given by the rotational part of the microdisplacement tensor (TEISSEYRE, 1995b). Moreover, in the symmetric micromorphic model of a near source zone, these rotational effects of seismic waves are related to microinertia tensor which describes inertia moments of microblocks forming a microstructure (TEISSEYRE, 1973a,b, 1974, 1986). Recently, TAKEO and ITO (1997) discussed the rotational effects by using the continuous dislocation theories which can be connected with the theory of micromorphic continuum (e.g., TEISSEYRE, 1973a,b, 1974, 1995b) and pointed out the possibility of estimating the rotational components.

Two examples in Figure 1 present those parts of the rotational seismogram derived from the azimuth system of seismographs which are contaminated by distinctly small errors (DROSTE and TEISSEYRE, 1976). Times of the P- and S-wave onsets are indicated by arrows. Epicentral distances here were extremely small. Rotational wave effects are very clearly observed close to the S-wave arrival.

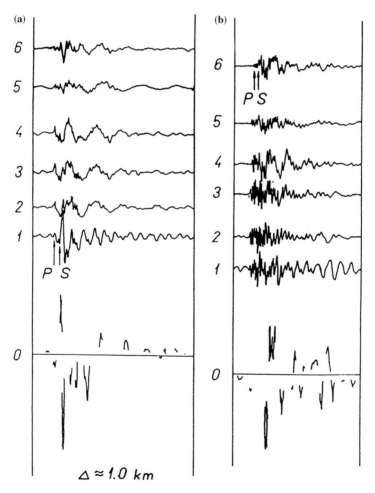

Figure 1
Six horizontal seismographs of azimuth components and the rotational seismograms in Silesia, Poland (after DROSTE and TEISSEYRE, 1976). (a) May 24, 1972, 09h 31m S2; (b) May 24, 1972, 12h 00m S2.

Equation of Equilibrium in Terms of Displacements: Navier Equation

Next we will focus on the equation of equilibrium in terms of displacements for the micromorphic theory of earthquakes (e.g., NAGAHAMA and TEISSEYRE, 1999). For a perfectly elastic isotropic material, the full stress-strain relations are

$$T_{ij} = \lambda e_{kk}\delta_{ij} + 2\mu e_{ij} \tag{8}$$

where T_{ij} is the stress tensor, λ and μ are the Lamé constants and $e_{kk} = e_{11} + e_{22} + e_{33}$ is the dilatation. Restricting ourselves to the solutions without the body forces, using eq. (8) and putting $T_{ij,j} = 0$, we can obtain the equilibrium equation in terms of displacements (the differential equation of static elasticity) in the form

$$\lambda\delta_{ij}\frac{\partial}{\partial x_j}e_{kk} + 2\mu\frac{\partial}{\partial x_j}e_{ij} = (\lambda + \mu)\frac{\partial^2}{\partial x_i \partial x_j}u_j + \mu\frac{\partial^2}{\partial x_j \partial x_j}u_i = 0. \tag{9}$$

Equation (9) is called Navier equation written in tensor notation for the elastic field (ENGLAND, 1971), which also can be derived from the first variation of the total elastic energy (LANCZOS, 1949).

For the symmetric micromorphic continuum (homogeneous microscopic continuum with spatially-independent material coefficients), the strain-stress relations and the expression for the energy density function can be written as

$$T_{kl} = \lambda e_{ii}\delta_{kl} + 2(\mu + \sigma)e_{kl} + \kappa\varepsilon_{kl} + \nu\varepsilon_{lk} \tag{10}$$

$$S_{kl} = \lambda e_{ii}\delta_{kl} + 2(\mu + 2\sigma)e_{kl} + (\kappa + \nu - \sigma)(\varepsilon_{kl} + \varepsilon_{lk}) \tag{11}$$

$$\Lambda_{plk} = 0, \tag{12}$$

where ν, κ, σ are material constants, $\mathbf{S}$ is the mocrostress tensor, Λ is the stress moment tensor and $\mathbf{T}$ is not necessarily symmetric for the symmetric micromorphic theory of the earthquakes (TEISSEYRE, 1973a,b, 1982, 1995b).

If the stress tensor $\mathbf{T}$ is symmetric ($\nu = \kappa$), from eqs (11–12) and the balance of moment of momentum, the microdisplacement tensor can be given by

$$\left(1 + \frac{N_{\langle kk \rangle}}{N_{\langle ll \rangle}}\right)\varphi_{lk} = \frac{\nu}{(\sigma - \nu)}\left(\frac{\partial u_l}{\partial x_k} + \frac{\partial u_k}{\partial x_l}\right). \tag{13}$$

Equation (13) is equivalent to the static case of the balance equations of moment of momentum for symmetric stress (see HANYGA and TEISSEYRE, 1973, 1974, 1975; TEISSEYRE, 1982, 1995b). From eq. (10), the equilibrium equation in terms of displacements is formally same as eq. (9) but with

$$\mu \to \mu + \nu + \sigma - \frac{2\nu^2}{(\nu - \sigma)\left(1 + \dfrac{N_{\langle kk \rangle}}{N_{\langle ll \rangle}}\right)}.$$

To describe the friction motion along a precut slip, we shall consider the ultrasmall rotating grains with their microinertia properties described by the tensor

$$N_{kl} = N\delta_{kl}. \tag{14}$$

In this case we will take the stress tensor **T** as asymmetric ($\nu \neq \kappa$) and the microdisplacement tensor can be expressed by

$$\varphi_{lk} = -\frac{\nu}{2(\hat{\nu} - \sigma)}\frac{\partial u_l}{\partial x_k} - \frac{\kappa}{2(\hat{\nu} - \sigma)}\frac{\partial u_k}{\partial x_l} \tag{15}$$

where $2\hat{\nu} = \nu + \kappa$. Equation (15) is equivalent to the static case of the balance equations of moment of momentum for asymmetric stress (see TEISSEYRE, 1995b). From eq. (10) the equilibrium equation in terms of displacements is formally the same as eq. (9) but with

$$\lambda \to \lambda + 2R + \frac{2\hat{\nu}R}{(\hat{\nu} - \sigma)}, \quad \mu \to \mu + \sigma + (\hat{\nu} - R) - \frac{\hat{\nu}(\hat{\nu} + R)}{(\hat{\nu} - \sigma)}$$

where $2R = \kappa - \nu$.

Thus for the micromorphic continuum for earthquakes, we can describe the equation of equilibrium in terms of displacements by the Navier equation in a classical theory of elasticity but we should use different Lamé coefficients. Here we can point out that the material constants of micromorphic continuum change with the deformation like self-organization.

Laplace Equations in Terms of Displacements

From the Navier equation one can easily obtain Laplace equations in terms of displacements. Differentiating both sides of eq. (9) with respect to x_i (which involves the results of separate differentiations for $i = 1, 2, 3$) and applying to both sides the operation *curl*, the following equations are obtained:

$$(\lambda + 2\mu)\nabla^2 \text{ div } \mathbf{u} = (\lambda + 2\mu)\nabla^2 e_{kk} = 0, \tag{16}$$

$$\mu\nabla^2\omega = 0, \tag{17}$$

where $\omega = \text{curl } \mathbf{u}$. Equations (16) and (17) are regarded as the kinematic compatibility conditions (LANDAU and LIFSHITZ, 1959a; COBBOLD, 1977; SORNETTE A. *et al.*, 1990; SORNETTE D. *et al.*, 1990).

Applying the Laplace operator to the Navier equation (9) yields

$$\Delta\Delta\mathbf{u} = 0 \tag{18}$$

which is called the biharmonic equation in displacements (LANDAU and LIFSHITZ, 1959a).

Particularly, in the absence of dilatations, i.e., in a pure shear field, eq. (9) is reduced to

$$\mu \nabla^2 \mathbf{u} = 0 \tag{19}$$

(LANDAU and LIFSHITZ, 1959a). This displacement field has been analyzed in detail by using the scalar potential ϕ from the view-point of the electric-elastic analogy (KNOPOFF, 1958).

Propagation of Deformation along Elastic Plate Boundaries Overlying a Viscoelastic Foundation: Macroscale Governing Equation

Based on a generalization of Elsasser and Rice's model of stress diffusion (ELSASSER, 1969; RICE, 1980), Rice and coworkers presented a simple two-dimensional model of a linearly elastic lithospheric plate of uniform thickness H overlying the (Maxwellian) viscoelastic asthenosphere (RICE, 1980; LEHNER et al., 1981; LI and RICE, 1983). When the equilibrium (the plane stress), stress-strain, and Maxwell coupling equations are combined, the thickness average displacement field $\mathbf{u}$ along the two-dimensional plate boundaries for the elastic plate is governed by

$$\frac{\partial u_r}{\partial t} = \left(\hbar + \lambda \frac{\partial}{\partial t} \right) \left\{ \frac{\partial^2 u_r}{\partial x_s \partial x_s} + \bar{\kappa} \frac{\partial^2 u_s}{\partial x_r \partial x_s} \right\}, \quad r, s = 1, 2 \tag{20}$$

$$\hbar \equiv hHG/\eta, \quad \lambda \equiv \zeta H, \quad \bar{\kappa} = \frac{1+\bar{\nu}}{1-\bar{\nu}}$$

where $\bar{\nu}$ is the Poisson ratio, h is the thickness of the asthenosphere, G is the shear modulus, η is the average viscosity of the asthenosphere and ζ is the effective length for short-time elastic coupling; x_1, x_2 are coordinates on the upper plate surface.

Differentiating both sides of eq. (20) with respect to x_r (which involves adding the results of separate differentiations for $r = 1, 2$), and applying to both sides the operation curl, the following equations are obtained (LEHNER et al., 1981):

$$\left(\hbar + \lambda \frac{\partial}{\partial t} \right) \nabla^2 e_{rr} = \{(1-\bar{\nu})/2\} \frac{\partial e_{rr}}{\partial t}, \tag{21}$$

$$\left(\hbar + \lambda \frac{\partial}{\partial t} \right) \nabla^2 \omega = \{(1-\bar{\nu})/2\} \frac{\partial \omega}{\partial t}. \tag{22}$$

In the case of non-elastic effects of the asthenosphere ($\lambda = 0$), i.e., in the case of the geological long-term deformation, eqs. (21) and (22) mean the diffusion equations of the isotropic strain/dilatation and the rigid-body rotation, respectively. Moreover, in this long-term case, eq. (20) can be reduced to the following macroscale diffusion equations (governing equations) of the average distortion of field β in the two-dimensional plate:

$$\frac{\partial \beta_{rs}}{\partial t} = h \left\{ (1 + \bar{\kappa}) \frac{\partial^2 \beta_{rs}}{\partial x_r \partial x_r} + \bar{\kappa} \frac{\partial^2 \beta_{sr}}{\partial x_s \partial x_s} \right\}. \tag{23}$$

Here we get for distortion the diffusion equations in a particular form without the random noise term of the nonlinear Langevin strain equation postulated by Sornette and coworkers (SORNETTE D. et al., 1990; SORNETTE and VIREUX, 1992; SORNETTE and SORNETTE, 1994). Equations (23) shows that the steady tectonic field within the lithospheric plates macroscopically obeys the steady flow ("flux") of the strains/distortions through the plate boundaries, which is similar to a strain input into the system from the outside.

In a static case such as freezing plate motions, eq. (20) can be reduced to the Navier equations of the average displacement fields as eqs. (9), and (21, 22) can be rewritten in the form of the Laplace equations (the static kinematic compatibility equations) for isotropic strain/dilatation and the rigid-body rotation similar to eqs. (16, 17), respectively.

Navier Equation, Laplace Field and Fractal Pattern Formation of Fracturing

Navier equation is a generalization of the Laplace equation which describes Laplacian fractal growth processes like diffusion limited aggregation (DLA), dielectric breakdown (DB) and viscous fingering in 2-D cells (LOUIS and GUINEA, 1987, 1989; FERNANDEZ et al., 1988). In particular, the fracturing fields have been studied as a scalar potential field by using the Laplace equation (19) both with the homogeneous elastic coefficients (e.g., TAGUCHI, 1989; STAKHOVSKY, 1995) and with the non-homogeneous elastic coefficients (e.g., TAKAYASU, 1985, 1986; SORNETTE and VANNESTE, 1996). However, we shall note an important difference with respect to DLA or DB, namely, the vectorial nature of the displacement field as compared to the scalar field of Laplacian fractal growth as DLA or DB.

The static kinematic compatibility conditions (eqs. 16, 17) derived from Navier equation (9) satisfy the generalized Laplace fields in terms of displacements. In particular, it has been often pointed out that the static kinematic compatibility conditions can be regarded as local diffusion-like conservation equations (Laplace equations) for strains (SORNETTE D. et al., 1990; SORNETTE A. et al., 1990). Moreover, the biharmonic equation (18) also suggests a close formal connection with DLA or DB (LOUIS and GUINEA, 1987, 1989; MEAKIN, 1991).

The vectorial nature of the displacement field and the fractal properties of fractures in the two-dimensional medium have been studied by using the discretization of Navier equation (LOUIS et al., 1986; LOUIS and GUINEA, 1987, 1989;

FERNANDEZ et al., 1988). Moreover, by regarding the fracturing in a three-dimensional medium as one in a "pseudo-two-dimensional + one-dimensional" medium, INAOKA and TAKAYASU (1996) recently succeeded in obtaining the more general solution of the Navier equation and pointed out that the size distribution of fragmentations can be expressed by

$$N(r) \propto r^{-D_S} \tag{24}$$

where $N(r)$ is the cumulative number of fragments larger than r and D_S is the fractal dimension on the size distribution. Then they derived $D_S = 2$ by following the stable-distribution theory (FELLER, 1966). This size distribution (eq. 24) means that fragmentation is statistically a scale-invariant process for the size distribution: self-similar for any size (e.g., MANDELBROT, 1982; TURCOTTE, 1986a; NAGAHAMA and YOSHII, 1994) and is often called the fractal size distribution (e.g., TURCOTTE, 1986a; NAGAHAMA and YOSHII, 1994).

Size Distributions of Fractures in the Lithosphere

The previous works have presented distributions of fracture size, such as length and width, on the ground surface. All of these distributions are power-law distributions as:

$$N(L) \propto L^{-D_F} \tag{25}$$

where $N(L)$ denotes the cumulative number of fractures on the ground surface larger than L, and D_F is a constant (e.g., NAGAHAMA, 1991, 1998; NAGAHAMA and YOSHII, 1994). Moreover, eq. (25) has also been traced to the size distribution of seismic faults by WENOUSKY et al. (1983).

According to the previous works on the power-law distribution of fracture sizes (e.g., NAGAHAMA, 1991, 1998), most D_F-values lie in the range $0 < D_F < 2$. NAGAHAMA (1991) and NAGAHAMA and YOSHII (1993, 1994) have also shown that the D_F-value reflects the rock properties (e.g., rock porosity and Weibull's uniformity coefficient of materials) and the tectonic conditions which should be related to the structural uniformity of the lithosphere.

Under many circumstances the cumulative number of earthquakes $N(M)$ with a magnitude greater than M satisfies the empirical Gutenberg-Richter's law (GUDENBERG and RICHTER, 1954)

$$\log N(M) = -bM + a \tag{26}$$

where a is a constant and b has been found to be very close to 1 (KANAMORI and ANDERSON, 1975). It has been shown that this frequency-magnitude relation for earthquake is equivalent to eq. (25), where the square root of the area of fault break corresponds to L (AKI, 1981; TURCOTTE, 1986a,b,c; NAGAHAMA, 1991).

MOGI (1962), in his experiments for elastic shocks accompanying fractures of various materials, pointed out that the b value increases with the degree of heterogeneity. Since MEREDITH and ATOKINSON (1983) pointed out that the b-value is related to the stress intensity factor, Main and coworkers (MAIN, 1991; MAIN et al., 1989; MEREDITH et al., 1990) have discussed the relationship between the stress intensity factor, b-value and the fractal dimensions concerning size distributions of fractures.

So, it is notable that the power-law form of eq. (25) holds over a wide range of fracture sizes in the lithosphere. This size distribution (25) is similar to the fractal size distribution of fragments (24). This suggests that eq. (25) is apparently valid for various fracture scales from a microcrack to a large fault, not only for the fault results spanning prolonged geological time but also for seismic faults (e.g., NAGAHAMA, 1991, 1998; NAGAHAMA and YOSHII, 1994). In other words, fracturing in the lithosphere also can be regarded as a scale-invariant process pertaining to the size distributions.

If the entire crust is fragmented and faults are the edges of these fragments (a current hypothesis being entertained by earth scientists; e.g., KING, 1983; TURCOTTE, 1986b), the log-linear frequency-magnitude distribution of earthquakes (eq. 26) is consistent with the fractal size distribution of fragmentation (eq. 24) and the b-values of eq. (26) are consistent with the negative exponent of a power-law distribution of fault lengths (e.g., AKI, 1981; TURCOTTE, 1986a,b,c; NAGAHAMA, 1991):

$$D_S = \frac{3b}{\delta} \qquad (27)$$

where δ is a constant which depends on the relative duration of the seismic source and the time constant of the recording system. Equation (27) indicates that the b-value is related to the fractal dimension of crustal fragmentation. From the experimental studies on tensile crack propagation for a variety of crystalline rocks, it follows that $\delta = 3.0$ is appropriate (MAIN et al., 1989), so that $D_S = b$ (MAIN et al., 1990). Moreover, for most earthquake studies $\delta = 1.5$ has been appropriate (KANAMORI and ANDERSON, 1975), so that $D_S = 2b$ (MAIN et al., 1989, 1990). However, DUBOLIS and NOVAILI (1989) obtained $\delta = 2.4$ from the seismicity of subduction zones between 100 and 700 km depth, so that $D_S = 1.25b$.

Utilizing the stable size distribution theory (FELLER, 1966), INAOKA and TAKAYASU (1996) derived $D_S = 2$ from their numerical analysis of Navier equation. Using their result and eq. (27), we can obtain $b = 1$ for most earthquake studies ($\delta = 1.5$; KANAMORI and ANDERSON, 1975).

Discussions

In the former sections, we mentioned the local conservation law (Laplace equation) for the strains within the lithosphere with microstructures and the macroscale average displacement field within the lithospheric plate overlying the asthenosphere. In this section, we will consider relations between them from the view-point of the theory of dissipative structures (non-equilibrium pattern formation theory) or the concept of the self-organized criticality (SOC).

In order to recognize the more physical meaning of the Navier equation (9) or the governing equation of plate motion (20), it is interesting to compare them with the Navier-Stokes equation or the phase diffusion equation. The Navier equation (9) is equivalent to the equation for the steady state flow of a viscous fluid at low Reynolds number (the Navier-Stokes equation; LANDAU and LIFSHITZ, 1959b). Equation (20) for the geological long-term deformation ($\lambda = 0$) is formally very similar to the phase diffusion equation in an isotropic two-dimensional medium for the non-equilibrium pattern formation:

$$\frac{\partial}{\partial t}\vec{\phi} = D^h_\perp \nabla^2 \vec{\phi} + (D^h_\parallel - D^h_\perp)\vec{\nabla}(\vec{\nabla}\vec{\phi}) \tag{28}$$

where $\vec{\phi}$ is the phase vector and $D^h_\perp$ and $D^h_\parallel$ are the parallel and transverse phase diffusion coefficients of the layered structure, respectively (e.g., WALGRAEF, 1988). By comparing eq. (28) with (20) in the case of $\lambda = 0$, the parameter $\bar{\kappa}$ of eq. (20) or the Lamé coefficients of the Navier equation (9) plays a role similar to the diffusion coefficients $D^h_\perp$ and $D^h_\parallel$. Similarly, CHELIDZE (1993) pointed out interesting possibilities following from the solution of the Navier equation, especially for the dynamic phase of crack development in terms of growth and diffusion models (STANLEY and OSTROVSKY, 1986).

The tectonic strains within the lithospheric plate are continuously produced from the plate boundaries by the tectonic loading and the strains are released by earthquakes (faulting). Moreover, NAGAHAMA (1994, 1996) pointed out the diffusional fractal character of earthquakes. Locally, the Navier equation or the kinematic compatibility conditions (16, 17) satisfy the diffusion-like conservation law for strains (the stationary strain balance) within the lithospheric plate with microstructures. Therefore, Sornette described these conservation equations as follows: these equations for strains can describe the existence of long-range correlations: for instance, once a fault is created the strain fields are redistributed over large distances (meaning algebraic decay of the strain) with strain enhancement at the tips; in other words, all past deformations in a given region within the lithospheric plate must be coherent with those of adjacent domains and with the offsets of both created and preexisting fracture (faults) (SORNETTE A. et al., 1990; see also SORNETTE D. et al., 1990). This local strain balance is quite similar to the hypothesis presented by ENYA (1901) that the main shock disturbs the strain distribution in the crust and aftershocks occur to decrease the heterogeneity of strain generated in the crust; in other words, the strain is in a balance/stationary state (see also ITO and MATSUZAKI, 1990; ITO, 1992).

Here we can see that there is an apparent paradox between the non-equilibrium macroscopic average flux of strains in the whole lithospheric system (eq. 20) and the local balance/stationary state of strains within the lithospheric plate with microstructures (eqs. 16 and 17). However, from the consideration of the self-organized criticality model (SOC model; BAK et al., 1987, 1988; ZHANG, 1989; HWA and KARDAR, 1989), if dynamics satisfies a local conservation law, then the steady configurations are ensured to be fractal or the system will be self-organized into a critical state. D. Sornette and coworkers also presented a similar SOC model for earthquakes which hypothesizes that the steady flow of tectonic stresses (strains) generates the fractal nature of earthquakes (SORNETTE D. et al., 1990; SORNETTE and VIREUX, 1992; SORNETTE and SORNETTE, 1994). Moreover, in another SOC model of earthquakes, ITO and MATSUZAKI (1990) took ENYA's (1901) idea that each earthquake distributes the strain field of the crust (local conservation rule) and derived some scaling laws of earthquakes. Therefore, it is concluded that a dynamic system with the non-equilibrium macroscopic flux of strains naturally evolves into a local balance/stationary state of strains with the scale-invariant properties of fractures in the lithosphere with microstructures.

Acknowledgements

We thank M. Kupková and two anonymous reviewers for insightful comments which improved the manuscript. The authors would like to thank Ms. A. Dziembowska for her improving the English of the manuscript. One of the authors (H. N) acknowledges the Scientist Exchange Research Fellowship between the Polish Academy of Sciences and the Japan Society for the Promotion of Science.

REFERENCES

AKI, K. A., *Probabilistic synthesis of precursory phenomena*. In *Earthquake Prediction: An International Review* (Maurice Ewing Series 4) (eds. Simpson, D. W., and Richards, P. G.) (Am. Geophys. Union, Washington D.C. 1981) pp. 566–574.

BAK, P., TANG, C., and WIESENFELD, K. (1987), *Self-organized Criticality: An Explanation of 1/f Noise*, Phys. Rev. Lett. *59*, 381–384.

BAK, P., TANG, C., and WIESENFELD, K. (1988), *Self-organized Criticality*, Phys. Rev. A *38*, 367–374.

BIELSKI, W., *Anisotropy in a micromorphic continuum*. In *Theory of Earthquake Premonitory and Fracture Processes* (ed. Teisseyre, R.) (Polish Scientific Publ., Warszawa 1995) pp. 633–639.

CHELIDZE, T. (1993), *Fractal Damage Mechanics of Geomaterials*, Terra Nova *5*, 421–437.

COBBOLD, P. R. (1977), *Compatibility Equations and the Integration of Finite Strains in Two Dimensions*, Tectonophys. *39*, T1–T6.

DROSTE, Z., and TEISSEYRE, R. (1976), *Rotational and Displacement Components of Ground Motion as Deduced from Data of the Azimuth System of Seismograph*, Publs. Inst. Geophys. Pol. Acad. Sci. *97*, 157–167 [in Polish with English abstract].

DUBOLIS, J., and NOVAILI, L. (1989), *Quantification of the Fracturing of the Slab Using a Fractal Approach*, Earth Planet. Sci. Lett. *94*, 97–108.

ELSASSER, W. M., *Convection and stress propagation in the upper mantle*. In *The Application of Modern Physics to the Earth and Planetary Interiors* (ed. Runcorn, S. K.) (Wiley Interscience, New York 1969) pp. 223–246.

ENGLAND, A. H., *Complex Variable Methods in Elasticity* (Wiley-Interscience, New York 1971).

ENYA, O. (1901), *On Aftershocks*, Rep. Earthq. Invest. Comm. *35*, 35–56 [In Japanese].

ERINGEN, A. C., *Theory of micropolar elasticity*. In *Fracture, vol. 2* (ed. Liebowitz, H.) (Academic Press, New York 1968) pp. 621–729.

ERINGEN, A. C., and SUHUBI, E. S. (1964), *Nonlinear Theory of Micro-elastic Solids. I*, Int. J. Eng. Sci. *2*, 189–203.

ERINGEN, A. C., and CLAUS, JR., W. D., *A micromorphic approach to dislocation theory and its relation to several existing theories*. In *Fundamental Aspects of Dislocation Theory* (eds. Simmons, J. A., deWit, R., and Bullough, R.) (Nat. Bur. Stand. (U.S.) Spec. Publ. 317, II 1970) pp. 1023–1040.

FELLER, W., *An Introduction to Probability Theory and its Applications, vol. 2* (Wiley, New York 1966).

FERNANDEZ, L., GUINEA, F., and LOUIS, E. (1988), *Random and Dendritic Patterns in Crack Propagation*, J. Phys. A: Math. Gen. *21*, L301–L305.

GUDENBERG, B., and RICHITER, C. F., *Seismicity of the Earth and Associated Phenomena*, 2nd ed. (Princeton Univ. Press, Princeton 1954).

HANYGA, A., and TEISSEYRE, R. (1973), *The Fundamental Source Solutions in the Symmetric Micromorphic Continuum*, Rivista Ital. Di Geofis. *22*, 336–340.

HANYGA, A., and TEISSEYRE, R. (1974), *Point Source Models in the Micromorphic Continuum*, Acta Geophys. Polon. *22*, 11–20.

HANYGA, A., and TEISSEYRE, R. (1975), *Linear Symmetric Micromorphic Thermoelasticity-source Solution and Wave Propagation*, Acta Geophys. Polon. *23*, 147–157.

HWA, T., and KARDAR, M. (1989), *Dissipative Transport in Open System: An Investigation of Self-organized Criticality*, Phys. Rev. Lett. *62*, 1813–1816.

IESAN, D. (1981), *Some Applications of Micropolar Mechanics to Earthquake Problems*, Int. J. Eng. Sci. *19*, 855–864.

INAOKA, H., and TAKAYASU, H. (1996), *Universal Fragment Size Distribution in a Numerical Model of Impact Fracture*, Physica A *229*, 5–25.

ITO, K. (1992), *Towards a New View of Earthquake Phenomena*, Pure appl. geophys. *138*, 531–548.

ITO, K., and MATSUZAKI, M. (1990), *Earthquake as Self-organized Critical Phenomena*, J. Geophys. Res. , B5 *95*, 6853–6860.

KANAMORI, H., and ANDERSON, D. L. (1975), *Theoretical Basis of Some Empirical Relations in Seismology*, Bull. Seismol. Soc. Am. *65*, 1073–1095.

KING, G. (1983), *The Accommodation of Large Strains in the Upper Lithosphere of the Earth and the Other Solids by Self-similar Fault Systems: The Geometric Origin of b-value*, Pure appl. geophys. *121*, 761–815.

KNOPOFF, L. (1958), *Energy Release in Earthquakes*, Geophys. J. *1*, 44–52.

LANCZOS, C., *The Variation Principles of Mechanics* (Univ. Toronto Press, Toronto 1949).

LANDAU, L. D., and LIFSHITZ, E. M., *Theory of Elasticity* (Pergamon Press, Oxford 1959a).

LANDAU, L. D., and LIFSHITZ, E. M., *Fluid Mechanics* (Pergamon Press, Oxford 1959b).

LEHNER, F. K., LI, V. C., and RICE, J. R. (1981), *Stress Diffusion along Rupturing Plate Boundaries*, J. Geophys. Res. *86*, 6155–6169.

LI, V. C., and RICE, J. R. (1983), *Preseismic Rupture Progression and Great Earthquake Instabilities at Plate Boundaries*, J. Geophys. Res. *88*, 4231–4246.

LOUIS, E., GUINEA, F., and FLORES, F., *The fractal nature of fracture*. In *Fractals in Physics* (eds. Pietronero, L., and Tosatti, E.) (Elsevier Sci. Pub., Amsterdam 1986) pp. 177–180.

LOUIS, E., and GUINEA, F. (1987), *The Fractal Nature of Fracture*, Europhys. Lett. *3*, 871–877.

LOUIS, E., and GUINEA, F. (1989), *Fracture as a Growth Process*, Physica D *38*, 235–241.

MAIN, I. G. (1991), *A Modified Griffith Criterion for the Evolution of Damage with a Fractal Distribution of Crack Lengths: Application to Seismic Event Rates and b-values*, Geophys. J. Int. *107*, 353–362.

MAIN, I. G., MEREDITH, P. G., and JONES, C. (1989), *A Reinterpretation of the Precursory Seismic b-value Anomaly from Fracture Mechanics*, Geophys. J. *96*, 131–138.

MAIN, I. G., PEACOCK, S., and MEREDITH, P. G. (1990), *Scattering Attenuation and the Fractal Geometry of Fracture Systems*, Pure appl. geophys. *133*, 283–304.

MANDELBROT, B. B., *The Fractal Geometry of Nature* (Freeman, New York 1982).

MEAKIN, P. (1991), *Models for Material Failure and Deformation*, Science *252*, 226–233.

MEREDITH, P. G., and ATOKINSON, B. K. (1983), *Stress Corrosion and Acoustic Emission during Tensile Crack Propagation in Whin Sill Dolerite and Other Basic Rocks*, Geophys. J. R. astr. Soc. *75*, 1–21.

MEREDITH, P. G., MAIN, I. G., and JONES, C. (1990), *Temporal Variations in Seismicity during Quasi-static and Dynamic Rock Failure*, Tectonophys. *175*, 249–268.

MOGI, K. (1962), *The Influence of the Dimensions of Specimens on the Fracture Strength of Rocks: Comparison between the Strength of Rock Specimens and that of the Earth's Crust*, Bull. Earthq. Res. Inst. *40*, 175–185.

NAGAHAMA, H. (1991), *Fracturing in the Solid Earth*, Sci. Repts. Tohoku Univ., 2nd ser. (Geol.) *61*, 103–126.

NAGAHAMA, H. (1994), *Self-affine Growth Pattern of Earthquake Rupture Zones*, Pure appl. geophys. *142*, 263–271.

NAGAHAMA, H. (1996), *Non-Riemannian and Fractal Geometries of Fracturing in Geomaterials*, Geol. Rund. *85*, 96–102.

NAGAHAMA, H. (1998), *Fractal Structural Geology*, Mem. Geol. Soc. Japan *50*, 13–19.

NAGAHAMA, H., and TEISSEYRE, R. (1999), *Micromorphic Continuum, Rotational Wave and Fractal Properties of Earthquakes and Faults*, Acta Geophys. Polon. *46*, 277–294.

NAGAHAMA, H., and YOSHII, K. (1993), *Fractal Dimension and Fracture of Brittle Rocks*, Int. J. Rock. Mech. Min. Sci. and Geomech. Abstr. *30*, 173–175.

NAGAHAMA, H., and YOSHII, K., *Scaling laws of fragmentation*. In *Fractal and Dynamical Systems in Geosciences* (ed. Kruhl, J. H.) (Springer-Verlag, Berlin 1994) pp. 25–36.

RICE, J. R., *The mechanics of earthquake rupture*. In *Physics of the Earth's Interior* (eds. Dziewonski, M. and Boschi, E.) (Italian Physical Society/North Holland, Amsterdam 1980) pp. 555–649.

SHIMBO, M. (1978), *A Geometrical Formation of Granular Media*, Theor. Appl. Mech. *26*, 473–480.

SORNETTE, D., and SORNETTE, A. (1994), *Comment on "On scaling relations for large earthquakes" by B. Romanowicz and J. B. Rundel, from the Perspective of a Recent Nonlinear Diffusion Equation Linking Short-time Deformation to Long-time Tectonics*, Bull. Seismol. Soc. Am. *84*, 1679–1683.

SORNETTE, D., and VIREUX, J. (1992), *Linking Short-timescale Deformation to Long-timescale Tectonics*, Nature *357*, 401–404.

SORNETTE, D., and VANNESTE, C. (1996), *Fault Self-organized by Repeated Earthquakes in a Quasi-static Antiplane*, Nonlinear Proc. Geophys. *3*, 1–12.

SORNETTE, D., DAVY, P., and SORNETTE, A. (1990), *Structuration of the Lithosphere in Plate Tectonics as a Self-organized Critical Phenomenon*, J. Geophys. Res. *95* (B11), 17353–17361.

SORNETTE, A., DAVY, P., and SORNETTE, D. (1990), *Growth of Fractal Fault Patterns*, Phys. Rev. Lett. *65*, 2266–2269.

STAKHOVSKY, I. R. (1995), *Fractal Geometry of Brittle Failure at Antiplanar Shear*, Phys. Solid Earth *31*, 268–275.

STANLEY, H., and OSTROVSKY, N., *On Growth and Form* (Nijhoff Publ., Amsterdam 1986).

SUHUBI, E. S., and ERINGEN, A. C. (1964), *Nonlinear Theory of Micro-elastic Solids. II*, Int. J. Eng. Sci. *2*, 389–404.

TAGUCHI, Y. (1989), *Fracture Propagation Governed by the Laplace Equation*, Phys. A *156*, 741–755.

TAKAYASU, H. (1985), *A Deterministic Model of Fracture*, Prog. Theor. Phys. *74*, 1343–1345.

TAKAYASU, H., *Pattern formation of dendritic fractals in fracture and electric breakdown*. In *Fractals in Physics* (eds. Pietronero, L., and Tosatti, E.) (Elsevier Sci. Publ. Amsterdam 1986) pp. 181–184.

TAKEO, M., and ITO, H. M. (1997), *What Can Be Learned from Rotational Motions Excited by Earthquakes?*, Geophys. J. Int. *129*, 319–329.

TEISSEYRE, R. (1973a), *Earthquake Processes in a Micromorphic Continuum*, Pure appl. geophys. *102*, 15–28.

TEISSEYRE, R. (1973b), *Earthquake Processes in a Micromorphic Continuum*, Rev. Roum. Géol. Géogr. Sér. de Géophys. *17*, 145–148.

TEISSEYRE, R., *Symmetric micromorphic continuum: wave propagation, point source solution and some applications to earthquake processes*. In *Continuum Mechanics Aspects of Geodynamics and Rock Fracture Mechanics* (ed. Thoft-Christensen) (D. Riedel Publ., Holland 1974) pp. 201–244.

TEISSEYRE, R. (1975), *On the Recent Development of Continuum Mechanics and its Application to Seismology*, Gerlands Beitr. Geophysik, Leipzig *84*, 501–508.

TEISSEYRE, R. (1978), *Relation Between the Defect Distribution and Stress. The Glacier Motion*, Acta Geophys. Polon. *26*, 283–290.

TEISSEYRE, R. (1982), *Some Seismic Phenomena in the Light of the Symmetric Micromorphic Theory*, J. Tech. Phys. *38*, 95–99.

TEISSEYRE, R., *Some problems of the continuum media and the applications to earthquake studies*. In *Continuum Theories in Solid Earth Physics* (ed. Teisseyre, R.) (Polish Scientific Publ., Warszawa-Elsevier, Amsterdam 1986) pp. 256–309.

TEISSEYRE, R., *Micromorphic model of a seismic source zone, 1. Introduction*. In *Theory of Earthquake Premonitory and Fracture Processes* (ed. Teisseyre, R.) (Polish Scientific Publ., Warszawa 1995a) pp. 613–615.

TEISSEYRE, R., *Micromorphic model of a seismic source zone, 2. Symmetric micromorphic theory; applications to seismology*. In *Theory of Earthquake Premonitory and Fracture Processes* (ed. Teisseyre, R.) (Polish Scientific Publ., Warszawa 1995b) pp. 616–627.

TEISSEYRE, R., and NAGAHAMA, H. (1999), *Micro-inertia Continuum: Rotations and Semi-waves*, Acta Geophys. Polon. *47*, 259–272.

TURCOTTE, D. L. (1986a), *Fractals and Fragmentation*, J. Geophys. Res. *91*, 1921–1926.

TURCOTTE, D. L. (1986b), *A Fractal Model for Crustal Deformation*, Tectonophys. *132*, 261–269.

TURCOTTE, D. L. (1986c), *Fractals in Geology and Geophysics*, Pure appl. geophys. *131*, 171–196.

TURCOTTE, D. L., *Fractals and Chaos in Geology and Geophysics* (Cambridge Univ. Press, Cambridge 1992).

WALGRAEF, D. (1988), *Instabilities and Patterns in Reaction-diffusion Dynamics*, Solid State Phenom. *384*, 77–96.

WENOUSKY, S. G., SCHOLZ, C. H., SHIMAZAKI, K., and MATSUDA, T. (1983), *Earthquake Frequency Distribution and the Mechanics of Faulting*, J. Geophys. Res. *88*, 9331–9340.

ZHANG, YI-C. (1989), *Scaling Theory of Self-organized Criticality*, Phys. Rev. Lett. *63*, 470–473.

(Received March 24, 1998, revised November 3, 1998, accepted December 3, 1998)

To access this journal online:
http://www.birkhauser.ch

Spatial Distribution of Aftershocks and the Fractal Structure of Active Fault Systems

KAZUYOSHI NANJO[1] and HIROYUKI NAGAHAMA[1]

Abstract—The relationship between the fractal dimensions of aftershock spatial distribution and of pre-existing fracture systems is examined. Fourteen main shocks occurring in Japan were followed by aftershocks, and the aftershocks occurred in swarms around the main shock. Epicentral distributions of the aftershocks exhibit fractal properties, and the fractal dimensions are estimated by using correlation integral. Observable pre-existing active fault systems in the fourteen aftershock regions have fractal structures, and the fractal dimensions are estimated by using the box-counting method. The estimated fractal dimensions derive positive correlation, showing independence from the main-shock magnitude. The correlation shows that aftershock distributions become less clustered with increasing fractal dimensions of the active fault system. That is, the clusters of the aftershocks are constrained under the fractal properties of the pre-existing active fault systems. If the fractal dimension of the active fault system is the upper limit value of the fractal dimension of the actual fracture geometries of rocks, then the clustering aftershocks manifest completely random and unpredictable distribution.

Key words: Aftershock, active fault system, fractal, fractal dimension, spatial distribution.

Introduction

Large earthquakes (main shocks) are mostly followed by aftershocks. Aftershock events occur in swarms around the main shock, called the aftershock region. Aftershock activities decay with time t, and the frequency of aftershocks $n(t)$ follows a power-law relationship (modified Omori formula: UTSU, 1961) with an exponent p close to 1, $n(t) \propto t^{-p}$. Size distribution of each aftershock sequence satisfies the Gutenberg-Richter relation (GUTENBERG and RICHTER, 1954), $\log N(M) = a - bM$, where $N(M)$ is the cumulative number of earthquakes with a magnitude greater than M, a is a constant and b has been found to be close to 1. Previous studies have reported significantly different values of p and b for various aftershock sequences worldwide, including aftershock sequences analyzed in this paper (e.g., UTSU, 1961; YOSHIDA and MIKAMI, 1986; KISSLINGER and JONES, 1991; GUO and OGATA, 1995, 1997; NANJO *et al.*, 1998).

---

[1] Institute of Geology and Paleontology, Graduate School of Science, Tohoku University, Sendai 980-8578, Japan, e-mail: nanjo@dges.tohoku.ac.jp

Analysis of spatial distributions of clustering aftershocks using the correlation integral (GRASSBERGER, 1983) exhibits a scale-invariant (fractal: MANDELBROT, 1983) behavior. The fractal dimensions of aftershock hypocentral distribution for Japan earthquakes have been reported ranging from 1.9 to 2.9 (YOSHIDA and MIKAMI, 1986; GUO and OGATA, 1995, 1997). YOSHIDA and MIKAMI (1986) analyzed the aftershock sequence of the 1984 Nagano-ken Seibu earthquake in Japan, and reported the fractal behavior of the time-spatial distributions. It was shown that the fractal properties were not changed by varying the lower limit of magnitude of aftershocks used in fractal analyses. Moreover, the decrease of irregularity of stress or strength in the lithosphere with increasing time after the main shock was discussed.

In the self-organized criticality model of earthquakes, ITO and MATSUZAKI (1990) took ENYA's (1901) idea, and derived successfully scaling laws of earthquakes mentioned above. ENYA's (1901) idea is that the main shock disturbs strain distribution, and the aftershocks occur to decrease the heterogeneity of the strain distribution in the crust. However, though the internal structures and discontinuities in the lithosphere seem to have an influence on the earthquakes, studies by ENYA (1901) and ITO and MATSUZAKI (1990) did not confirm their influence. NAGAHAMA and TEISSEYRE (1998, 2000, this issue) discuss the relationship between the fractal properties of fracturing in the lithosphere and of the micromorphic continuum which is the continuum with microstructure. Therefore, we need to explore the relationship between the fractal properties of earthquakes and of the discontinuities in the lithosphere. One approach to look into the relationship is to examine the relationships among seismic parameters and the fractal dimension of the fracture system in the lithosphere.

A theoretical study derived the modified Omori formula from a preliminary statistical model where aftershocks are produced by a random walk on a pre-existing fracture system (HIRATA, 1986). The derived result is the direct connection between p and the fractal dimension of the pre-existing fracture system. This study implies that fractal properties of aftershocks are constrained by the fractal structure of the pre-existing fracture system. If some relationships among the seismic parameters of aftershocks and the fractal dimension of the pre-existing fracture system are found, then the correctness of the implication of the model is suggested.

To this end, NANJO et al. (1998) examined the relationships among p, b and the fractal dimension of pre-existing fault systems. p and b-values were estimated for 15 aftershock sequences in Japan (Figs. 1 and 2; Table 1). The fractal dimensions (capacity dimensions) D_0 of the pre-existing fracture system were estimated for pre-existing active fault systems in their aftershock regions by using the box-counting method (Fig. 3; Table 1). They found negative correlations of p to D_0 and b. The correlations demonstrate the correctness of the implication from the model (HIRATA, 1986). However, the relationship between the fractal dimension of

Table 1

Earthquake origins of 15 main shocks followed by analyzed aftershock sequences and estimated values of p, b, D_0 and D_2. Loc: location (north latitude, east longitude) (deg.); Dep: focal depth (km); M: magnitude. Main data source of the earthquakes is from 'SEIS-PC' (ISHIKAWA, 1986; ISHIKAWA et al., 1989). Earthquake data source of the Hyogo-ken Nanbu earthquake is from the Disaster Prevention Research Institute of Kyoto University. Data on active faults are from 'Active Faults in Japan' (THE RESEARCH GROUP FOR ACTIVE FAULTS OF JAPAN, 1991). Some older earthquakes have a lower reliability of the depth value, presented by '–' in the column Dep. Values in the parentheses are error bounds. For spatial distribution, Central Gifu prefecture and Eastern Yamanashi prefecture do not show scale invariance, and the Hyogo-ken Nanbu earthquake is not used in this paper, indicated by '–' in the column D_2

	Date	Loc	Dep	M	p	b	D_0	D_2
Nishisaitama earthquake	1931.9.21	(36.2, 139.1)	–	6.9	1.13 (0.02)	0.85 (0.01)	1.09 (0.01)	1.70 (0.01)
Tottori earthquake	1943.9.10	(35.5, 134.1)	–	7.2	1.19 (0.04)	0.79 (0.01)	0.91 (0.01)	1.82 (0.01)
Mikawa earthquake	1945.1.13	(34.7, 137.1)	–	6.8	1.08 (0.04)	0.89 (0.01)	1.18 (0.01)	1.75 (0.00)
Fukui earthquake	1948.6.28	(36.2, 136.2)	–	7.1	1.07 (0.02)	0.85 (0.00)	1.32 (0.01)	1.82 (0.00)
Imaichi earthquake	1949.12.26	(36.6, 139.8)	–	6.4	1.10 (0.02)	0.91 (0.05)	1.14 (0.01)	1.69 (0.01)
Kitamino earthquake	1961.8.19	(36.0, 136.8)	–	7.0	1.13 (0.04)	0.65 (0.03)	1.22 (0.01)	1.79 (0.01)
Central Gifu prefecture	1969.9.9	(35.8, 137.1)	–	6.6	1.08 (0.03)	0.87 (0.01)	1.21 (0.01)	–
Eastern Yamanashi prefecture	1976.6.16	(35.5, 139.0)	20	5.5	1.06 (0.01)	0.89 (0.01)	1.30 (0.01)	–
Boundary between Kanagawa and Yamanashi prefectures	1983.8.8	(35.5, 139.0)	22	6.0	1.06 (0.04)	0.87 (0.00)	1.28 (0.01)	1.84 (0.00)
Eastern Tottori prefecture	1983.10.31	(35.4, 133.9)	15	6.2	1.20 (0.05)	0.77 (0.01)	0.86 (0.00)	1.75 (0.01)
Near Unzen	1984.8.6	(32.8, 130.2)	7	5.7	0.93 (0.03)	1.30 (0.01)	1.39 (0.01)	1.91 (0.00)
Nagano-ken Seibu earthquake	1984.9.14	(35.8, 137.6)	2	6.8	1.01 (0.02)	0.97 (0.00)	1.45 (0.01)	1.94 (0.00)
Northern Nagano prefecture	1986.12.30	(36.6, 137.9)	3	5.9	1.18 (0.07)	0.64 (0.01)	1.17 (0.01)	1.77 (0.00)
Yamaguchi prefecture	1987.11.18	(34.1, 131.5)	8	5.2	1.21 (0.02)	0.59 (0.03)	1.08 (0.00)	1.68 (0.00)
Hyogo-ken Nanbu earthquake	1995.1.17	(34.6, 135.0)	18	7.2	0.67 (0.01)	0.63 (0.00)	1.48 (0.00)	–

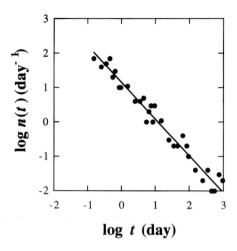

Figure 1
Aftershock decay of the Fukui earthquake. Aftershock frequency per day $n(t)$ (day^{-1}) is plotted against time t (day) on the double logarithmic graph. Aftershock frequency statistically obeys a power law (the modified Omori formula) with $p = 1.07 \pm 0.02$, the slope of the least-square regression line fitted to the data in the range $0.1 \leq t \leq 1000$.

aftershock spatial distribution and the fractal dimension of the active fault system has not been examined. If relationships between the fractal dimensions are found, then this supports the hypothesis of HIRATA (1986).

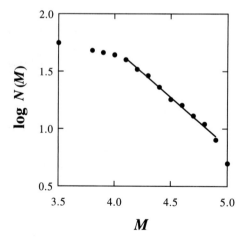

Figure 2
Size distribution of the Fukui earthquake. The number of earthquakes $N(M)$ with magnitudes greater than M is plotted against M on the semi-logarithmic graph. Because we observe that at $M = 4.1$, the cumulative number of small events begins to fall off the Gutenberg-Richter relation (MALIN et al., 1989), $b = 0.85 \pm 0.00$ of the Gutenberg-Richter relation is estimated from the slope of the least-square regression line fitted to the data in the range $4.1 \leq M \leq 4.9$.

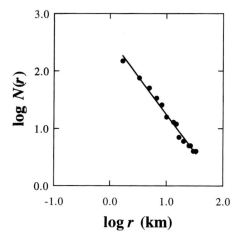

Figure 3
Results of fractal analysis of the active fault system in the aftershock region of the Fukui earthquake. The number of boxes entered by the line of active faults $N(r)$ is plotted against the side length of square box r (km). The estimated fractal dimension (D_0) is 1.32 ± 0.01, the least-square regression line fitted to these data in the range $1.7 \leq r \leq 33.3$.

Here, the fractal dimensions of aftershock spatial distributions are estimated for epicentral distribution of 14 aftershock sequences following 14 large events in Japan (Table 1). The fractal dimensions of the pre-existing fracture system are estimated for observable active fault systems in the 14 aftershock regions. The relationship between the above two sets of fractal dimensions is examined. Finally, fractal properties of aftershocks are discussed.

Data

In this paper, 14 aftershock sequences in Japan, and active fault systems in the 14 aftershock regions are analyzed (Table 1). The data used are fundamentally the same as those used by NANJO et al. (1998), however the data of the Disaster Prevention Research Institute of Kyoto University are not used.

The data source of the earthquakes is the file 'SEIS-PC' (ISHIKAWA, 1986; ISHIKAWA et al., 1989), in which data of earthquakes occurring in and around Japan since 1885 are listed (original data are from the Japan Meteorological Agency). Earthquakes with $M \geq 0$ are treated in our analyses. $M = 0$ is the lowest magnitude in the data set.

The data on active faults are from 'Active Faults in Japan' (THE RESEARCH GROUP FOR ACTIVE FAULTS OF JAPAN, 1991) which consists of map sheets of active faults. In this book, active faults are defined as faults which were active during Quarternary (2 Ma to the present) and may possibly move again soon. The listed active faults on land were identified by careful interpretation of air photographs and were confirmed by land survey. This research group classified all faults mapped in this book into three groups on the basis of the judgements as follows: (1) it is certain beyond doubt that the fault was active during the Quarternary; (2) it is not definitely certain that the fault was active during the Quarternary but it is possible to infer the sense of the displacement; and (3) a fault is a lineament (linear topography) suspected of being a fault active during the Quarternary. Irrespective of the classification, all active faults listed up were used in our analyses. Active submarine faults identified by seismic reflection profiles were also included when they are within the aftershock regions.

Definitions of Aftershocks and Aftershock Region

Aftershocks must be clearly distinguished from other background events. It is a serious problem in that the definition of an aftershock varies from one investigator to the other (FROHLICH, 1989; KISSLINGER, 1996). Following UTSU (1991), we define aftershock events as earthquakes in swarms around the main shock (large event), tentatively within the next 1000 days (NANJO et al., 1998). For the convenience of treatment, the aftershock region is re-defined as the smallest rectangle in which all aftershock epicenters are included (NANJO et al., 1998). Figure 4 shows the spatial epicentral distribution (solid circles) of the Fukui earthquake. Figure 5 shows the aftershock region (surrounded by broken lines) of the Fukui earthquake and the active faults (solid lines) in and around it.

Two Point-correlation Integral and Spatial Distribution of Aftershocks

Spatial distributions of given earthquakes (e.g., regional seismicity and aftershocks) and of acoustic emissions (AE) in rock specimen show scale-invariant behavior (e.g., KAGAN and KNOPOFF, 1978, 1980; SADOVISKY et al., 1984; YOSHIDA and MIKAMI, 1986; HIRATA, 1989a; HIRATA et al., 1987; GUO and OGATA, 1995, 1997; ONCEL et al., 1995). The two-point correlation integral $C(l)$ is often used to evaluate the structure of a point collection (e.g., earthquakes, AE and galaxy) distributed in a space (MANDELBROT, 1983). $C(l)$ is defined as:

$$C(l) = \frac{1}{N^2} \sum_{i \neq j} H(l - L) \tag{1}$$

where N is the total number of points (earthquakes), $L = \|\bar{x}_i - \bar{x}_j\|$ is a distance between position vectors $\bar{x}_i$ and $\bar{x}_j$ of points, and $H(z)$ is the Heaviside function, which is 1 when $z \geq 0$, and is 0 when $z < 0$. If a point collection has scale-invariance, then $C(l)$ is represented by a power law form, given by:

$$C(l) \propto l^{D_2} \qquad (2)$$

where D_2 is defined as the fractal dimension (correlation dimension), which gives the lower limit of the Hausdorff dimension (e.g., GRASSBERGER, 1983; MANDELBROT, 1983). D_2 provides a measure of the degree of fractal clustering of points in a space, with lower fractal dimensions indicating tighter clusters (HIRATA et al., 1987). If the point distribution is completely random and unpredictable in a two-dimensional space, D_2 is 2. We use $C(l)$ to obtain D_2. In the present work, D_2 is given by the slope of the least-square regression line fitted to data which are visibly in a straight line within an observed range on the double logarithmic graph, though ENEVA (1996) utilized a different technique to determine the straight range to use in estimating the correlation dimensions.

In this study, $C(l)$ is evaluated for epicentral distribution of aftershocks. To evaluate $C(l)$ of the epicentral distribution, L between two given earthquakes must be defined. We assume that epicenters lie on the surface on an unit sphere. L (degree) is calculated by using a spherical triangle (HIRATA, 1989a; ONCEL et al., 1995), given by:

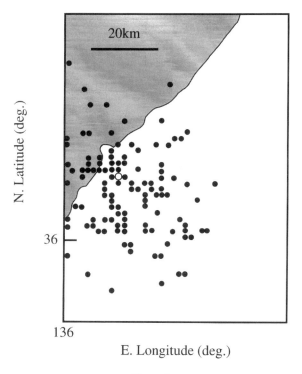

Figure 4
Epicentral distribution of aftershocks following the Fukui earthquake. Open and solid circles are epicenters of the main shock and of aftershocks, respectively.

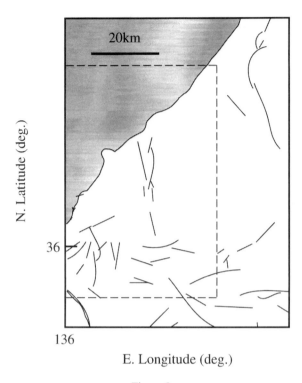

Figure 5
Active faults in and around the aftershock region of the Fukui earthquake. Solid curves are active faults. The aftershock region is the rectangle surrounded by broken lines.

$$L = \cos^{-1}\{\cos\theta_1 \cos\theta_2 + \sin\theta_1 \sin\theta_2 \cos(\omega_1 - \omega_2)\} \quad (3)$$

where (θ_1, ω_1) and (θ_2, ω_2) are north latitude (deg.) and east longitude (deg.) of two given epicenters.

Figure 6 shows $C(l)$ as a function l (deg.) for the Fukui earthquake. Equation (2) holds statistically in the range $-3.4 \leq \log(l) \leq -2.4$ with $D_2 = 1.82 \pm 0.00$. (2) holds statistically for the other 11 aftershock sequences. Estimated values of D_2 range from 1.68 to 1.94 (Table 1). The other two aftershock sequences (Central Gifu prefecture and Eastern Yamanashi prefecture) do not show the power-law relation (equation (2)). Therefore, D_2-values are not estimated for the two aftershock sequences.

Box-counting Method

Fracture systems including fault systems show a statistical self-similarity over a wide range of size scales (e.g., TURCOTTE, 1992). As a consequence, fracture systems are characterized by a power-law, with a characteristic exponent called a fractal dimension. Box-counting procedures were used to demonstrate the self-similarity of fault geometry (e.g., HIRATA, 1989b; NANJO et al., 1998). Fractal dimension measured by the box-counting method is called a 'box-count dimension' in a practical sense, and is equivalent to a 'capacity dimension' in a mathematical sense, usually denoted by D_0 (e.g., KORVIN, 1992). D_0 can be used to quantify the occupancy rate of a fractal geometrical pattern in a space. Therefore, D_0 estimated for a fracture system characterizes the roughness of fracturing or the occupancy of the fracture system (NANJO et al., 1998). In this study, the box-counting method is used to obtain D_0. The box-counting procedure used is given below:

An active fault system in a given extent (aftershock region in this study) is overlaid with a grid of square boxes; grids of different size boxes are used. The aftershock region thus must be rectangle. The number of boxes $N(r)$ of size r required to cover the fault system is plotted on the double logarithmic graph as a function of r. If the active fault system has a self-similar structure, then we derive the following relationship,

$$N(r) \propto r^{-D_0} \quad (4)$$

where D_0 is defined as fractal dimension (box-counting dimension or capacity dimension). In this study D_0 is given by the slope of the least-square regression line fitted to data within an effective range on the double logarithmic graph.

Figure 3 shows results of the fractal analysis of the active fault system in the aftershock region of the Fukui earthquake. Equation (4) holds statistically in the range $1.7 \leq r \leq 33.3$ (km) with $D_0 = 1.09 \pm 0.01$. (4) holds statistically for the other 13 active fault systems. Estimated values of D_0 range from 0.86 to 1.48 (Table 1).

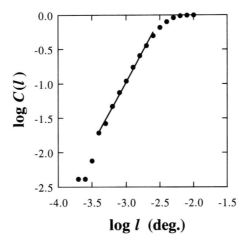

Figure 6
Correlation integral $C(l)$ as a function distance l (deg.) for the Fukui earthquake. The estimated fractal dimension (D_2) is 1.82 ± 0.00, the slope of the least-square regression line fitted to these data in the range $-3.4 \leq \log(l) \leq -2.4$.

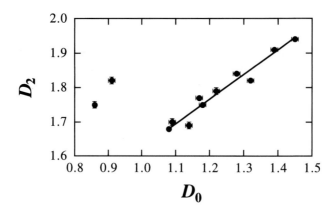

Figure 7
The fractal dimensions (D_0) of the active fault system vs. the fractal dimensions (D_2) of epicentral distribution. Excepting the Tottori earthquake and the Eastern Tottori prefecture, positive correlation is found with the least-square regression line, $D_2 = (0.72 \pm 0.02)D_0 + (0.91 \pm 0.02)$.

Relationship between D_2 and D_0

Figure 7 illustrates the relationship between D_2 and D_0. Excluding the points of the Tottori earthquake and the eastern Tottori prefecture, a positive correlation is found with a least-square regression line given by

$$D_2 = (0.72 \pm 0.02)D_0 + (0.91 \pm 0.02). \tag{5}$$

The two aftershock sequences were excepted from the regression because the two points were far from the general tendency and the D_2-values have lower reliability than the others. The number of aftershocks in the excepted cases was not large enough to estimate D_2-values. The two aftershock sequences consist of the smallest and the second smallest numbers in the 12 aftershock sequences. This positive correlation indicates that the fractal clustering of aftershock distribution D_2 becomes less strong with increasing D_0. Estimated D_2-values were less than 2 (Table 1). Moreover, D_2-values which are given by substituting estimated D_0-values (Table 1) into (5) are less than 2. Therefore, (5) means that the aftershocks tend to distribute more randomly and unpredictably with increasing occupancy of the active fault systems in the aftershock regions.

Whether or not the correlation between D_2 and D_0 depends on the magnitude of the main shock (Table 1) is looked into. Here the relationships among them are examined. No correlations among them are found. No correlations indicate independence of the correlation between D_2 and D_0 from the main-shock magnitude. Therefore the correlation between D_2 and D_0 is a scale-invariant relationship between aftershock distribution and the active fault system.

Discussion

The estimated values of the box-counting dimension of the active fault systems, D_0, ranged from 0.86 to 1.48 (Table 1). The upper limits of D_0 of rock fracture geometry were reported to be nearly equal to 1.5 (HIRATA, 1989b; NANJO et al., 1998). Here note that the maximum value of D_0 in this paper is quite close to the upper limit of the fractal dimension $D_0 = 1.5$. If D_0 of the active fault system in this paper is equal to the upper limit of $D_0 = 1.5$, then D_2 is given to be quite close to 2.0 by (5). Actually, the maximum value of the estimated D_2 is 1.94, which is close to 2.0. In the case of $D_2 = 2.0$ in a two-dimensional space, the clustering aftershocks exhibit completely random and unpredictable distribution. However, in observed nature, when the estimated values of D_0 range from 0.86 to 1.48, the estimated values of D_2 range from 1.68 to 1.94 (Table 1). That is, each cluster of the aftershocks shows the fractal properties under the constraints of the fractal structure of the active fault systems.

HIRATA (1986) derived the modified Omori formula from a model in which aftershocks are produced by a random walk on a pre-existing fracture system, resulting in the direct connection between p and the fractal dimension of the pre-existing fracture system. This model implies that fractal properties of aftershocks are constrained by the fractal structure of the pre-existing fracture system. NANJO et al. (1998) derived the correlations among p, b and D_0 from analyzing the natural observable data of seismicity and active faults. Therefore equation (5) and the correlations in NANJO et al. (1998) support the correctness of the implication from the model.

MOGI (1962, 1967) has discussed the relationship between aftershock activity and the crustal heterogeneity caused by faulting. YOSHIDA and MIKAMI (1986) discussed the decrease of irregularities of stress or strength in the lithosphere with increasing time after the main shock, from the fractal analyses of the aftershock sequence of the 1984 Nagano-ken Seibu earthquake in Japan. MIKUMO and MIYATAKE (1979) took a similar idea in the frictional fault model with nonuniform strength and presented a negative relationship between p and b by their simulation. The negative correlation between p and b in NANJO et al. (1998) supports qualitatively the frictional fault model with nonuniform strength.

ENYA (1901) hypothesized that the main shock disturbs strain distribution and the aftershocks occur to decrease the heterogeneity of the strain distribution in the crust. In the self-organized criticality model of earthquakes, ITO and MATSUZAKI (1990) took ENYA's (1901) idea, and successfully derived scaling laws of earthquakes. However, these studies have not established the relationship between the fractal properties of the aftershocks and the pre-existing discontinuities in the lithosphere.

In general, the internal structures and the discontinuities in the lithosphere seem to influence the earthquakes. NAGAHAMA and TEISSEYRE (1998, 2000, this issue) discuss the fractal properties of fracturing such as faulting or earthquakes in the lithosphere from the point of view of the continuum with microstructure (the micromorphic continuum). The pre-existing active faults analyzed in NANJO et al. (1998) and this paper can be regarded as the discontinuities in the lithosphere. NANJO et al. (1998) and this paper establish that the structures of active fault systems essentially influence the fractal properties of the aftershocks. By comparing our results with previous studies, the crustal heterogeneity (MOGI, 1962, 1967), the irregularities of stress or strength in the lithosphere (MIKUMO and MIYATAKE, 1979; YOSHIDA and MIKAMI, 1986) and the heterogeneity of the strain distribution in the crust (ENYA, 1901) may be regarded as the fractal structures of active fault systems.

Very often, special cluster patterns like doughnut patterns of the foreshocks have been pointed out (e.g., MOGI, 1985). Considering our results, we note that the foreshock cluster pattern may be related to the fractal structures of the active fault systems in which the earthquakes occur. To examine the relationship for the practical purpose of the prediction of large earthquakes requires further research.

Conclusions

Epicentral distributions of the aftershocks following 14 main shocks which occurred' in Japan, exhibit fractal properties. The fractal dimensions, D_2, range from 1.61 to 1.94. Moreover, observable pre-existing active fault systems in the 14 aftershock regions exhibit fractal behavior, and the fractal dimensions, D_0, range from 0.86 to 1.48. The relationship between D_2 and D_0, which excludes the main shock magnitude, is expressed as $D_2 = (0.72 \pm 0.02)D_0 + (0.91 \pm 0.02)$, and implies that the aftershock distribution tends to cluster more loosely with an increase in the occupancy of the active faults in the aftershock region. If the fractal dimension of the active fault system is the upper limit value of the fractal dimension of the actual fracture geometries in rocks, then the clustering aftershocks display completely random and unpredictable distribution. Therefore it is concluded that each cluster of the aftershocks reveals the fractal properties under the constraints of the fractal structure of the active fault systems.

Acknowledgements

We thank M. Satomura, Y. Ogata, T. Engelder, K. Ito, I. G. Main and K. Otsuki for their valuable discussions of our earlier research (NANJO et al., 1998), and also extend gratitude to M. Kupkova and two anonymous reviewers for their helpful and valuable suggestions which improved our earlier manuscript and for revising the English style of the earlier version manuscript. We are grateful to Y. Ishikawa for permission to use the file 'SEIS-PC'.

References

Eneva, M. (1996), *Effect of Limited Data Sets in Evaluating the Scaling Properties of Spatially Distributed Data: An Example from Mining-induced Seismic*, Geophys. J. Int. *124*, 773–786.
Enya, O. (1901), *On Aftershocks*, Rep. Earthq. Inv. Comm. *35*, 35–56 (in Japanese).
Frohlich, C. (1989), *The Nature of Deep-focus Earthquake*, Annu. Rev. Earth Planet. Sci. *17*, 227–254.
Grassberger, P. (1983), *Generalized Distributions of Strange Attractors*, Phys. Lett. *97*, 227–230.
Guo, Z., and Ogata, Y. (1995), *Correlation between Characteristic Parameters of Aftershock Distributions in Time, Space and Magnitude*, Geophys. Res. Lett. *22*, 993–996.
Guo, Z., and Ogata, Y. (1997), *Statistical Relations between the Parameters of Aftershocks in Time, Space and Magnitude*, J. Geophys. Res. *102*, 2857–2873.
Gutenberg, B., and Richter, C. F., *Seismicity of the Earth* (Princeton Univ. Press, Princeton 1954).
Hirata, T. (1986), *Omori's Power Law for Aftershocks and Fractal Geometry of Multiple Fault System*, Zishin (J. Seismol. Soc. Japan) *39*, 478–481 (in Japanese).
Hirata, T. (1989a), *A Correlation between b Value and the Fractal Dimension of Earthquakes*, J. Geophys. Res. *94*, 7507–7514.
Hirata, T. (1989b), *Fractal Dimension of Fault Systems in Japan, Fractal Structure in Rock Fracture Geometry at Various Scales*, Pure appl. geophys. *131*, 157–170.
Hirata, T., Satoh, T., and Ito, K. (1987), *Fractal Structure of Spatial Distribution of Microfracturing in Rock*, Geophys. J. R. Astron. Soc. *90*, 369–374.
Ishikawa, Y. (1986), *SEIS-PC, New Version*, Geol. Data Processing *11*, 65–74.
Ishikawa, Y., Mochizuki, E., Sakuma, K., Yamamoto, M., and Naruto, N. (1989), *Compile of Focal Mechanism Solutions*, Gekkan Chikyu (Earth Monthly) *3*, 219–223 (in Japanese).
Ito, K., and Matsuzaki, M. (1990), *Earthquakes as Self-organized Critical Phenomena*, J. Geophys. Res. *65*, 6853–6860.
Kagan, Y., and Knopoff, L. (1978), *Statistical Study of the Occurrence of Shallow Earthquakes*, Geophys. J. R. Astron. Soc. *55*, 67–86.
Kagan, Y., and Knopoff, L. (1980), *Spatial Distribution of Earthquakes, the Two-point Correlation Function*, Geophys. J. R. Astron. Soc. *62*, 303–320.
Kisslinger, C. (1996), *Aftershocks and Fault-zone Properties*, Advs. Geophys. *38*, 1–36.
Kisslinger, C., and Jones, L. M. (1991), *Properties of Aftershocks in Southern California*, J. Geophys. Res. *96*, 11,947–11,958.
Korvin, G., *Fractal Models in the Earth Sciences* (Elsevier, Amsterdam 1992).
Malin, P. E., Blakeslee, S. N., Alvarez, M. G., and Martin, A. J. (1989), *Microearthquake Imaging of the Parkfield Asperity*, Science *244*, 557–559.
Mandelbrot, B. B., *The Fractal Geometry of Nature* (W.H. Freeman and Company, New York 1983).
Mikumo, T., and Miyatake, T. (1979), *Earthquake Sequences on Frictional Fault Model with Non-uniform Strengths and Relaxation Times*, Geophys. J. R. Astron. Soc. *59*, 497–522.
Mogi, K. (1962), *On the Time Distribution of Aftershocks Accompanying the Recent Major Earthquakes in and near Japan*, Bull. Earthq. Res. Inst. *40*, 107–124.
Mogi, K. (1967), *Regional Variation of Aftershock Activity*, Bull. Earthq. Res. Inst. *45*, 711–725.
Mogi, K., *Earthquake Prediction* (Academic Press, Tokyo 1985).
Nagahama, H., and Teisseyre, R. (1998), *Micromorphic Continuum, Rotational Wave and Fractal Properties of Earthquakes and Faults*, Acta Geophys. Pol. *46*, 277–294.
Nagahama, H., and Teisseyre, R. (2000), *Micromorphic Continuum and Fractal Fracturing in the Lithosphere*, Pure appl. geophys. (this volume).
Nanjo, K., Nagahama, H., and Satomura, M. (1998), *Rates of Aftershock Decay and the Fractal Structure of Active Fault Systems*, Tectonophysics *287*, 173–186.
Oncel, A. O., Alptekin, O., and Main, I. G. (1995), *Temporal Variations of the Fractal Properties of Seismicity in the Western Part of the North Anatolian Fault Zone: Possible Artifacts Due to Improvements in Station Coverage*, Nonlinear Processes Geophys. *2*, 145–157.
Sadovsky, M. A., Golubeva, T. V., Pisarenko, V. F., and Shnirman, M. G. (1984), *Characteristic Dimension of Rock and Hierarchical Properties of Seismicity*, Izv. Acad. Sci. USSR Phys. Solid Earth, Engl. Transl. *20*, 87–96.

THE RESEARCH GROUP FOR ACTIVE FAULTS OF JAPAN, *Active Faults in Japan, Sheet Maps and Inventories* (revised edition) (Univ. of Tokyo Press, Tokyo 1991) (in Japanese with English summary).
TURCOTTE, D. L., *Fractals and Chaos in Geology and Geophysics* (Cambridge Univ. Press, New York 1992).
UTSU, T. (1961), *A Statistical Study on the Occurrence of Aftershocks*, Geophys. Mag. *30*, 521–605.
UTSU, T., *Seismology*, 2nd ed. (Kyoritsu Shuppan, Tokyo 1991) (in Japanese).
YOSHIDA, A., and MIKAMI, N., *On the time-spatial distribution of aftershocks-introduction*. In *Mathematical Seismology* (ed. M. Saito) (Inst. Statist. Math., Tokyo 1986) pp. 98–108 (in Japanese).

(Received March 31, 1998, revised January 20, 1999, accepted January 21, 1999)

To access this journal online:
http://www.birkhauser.ch

Scale-invariance Properties in Seismicity of Southern Apennine Chain (Italy)

V. LAPENNA,[1] M. MACCHIATO,[2] S. PISCITELLI[1] and L. TELESCA[1]

Abstract—Advanced statistical methodologies have been applied to detect temporal and spatial scaling-invariance laws in seismicity observed in a seismogenetic area of the Southern Apennine chain, which in past and recent years was devastated by destructive events. A deeper analysis of the seismicity observed in this area during the period 1983–1995 is presented. A counting statistical procedure has been applied to highlight scaling laws in the time domain and a fractal approach has been carried out to reveal spatial self-similarity. The main features of the dynamic system under study have been depicted using the key parameters of the more recurrent scaling laws observed in the spatial and temporal domains and we point out that the seismotectonic environment of the Southern Apennine chain can be considered as an extended dissipative system. The comparative analysis of different dynamical parameters of earthquake sequences provides a more comprehensive evaluation of seismotectonic processes in the investigated area.

Key words: Seismicity, fractal analysis, counting statistics.

1. Introduction

Today it is widely accepted that many spatially extended dissipative dynamical systems, that are systems with both many temporal and spatial degrees of freedom, evolve toward a critical state, with no intrinsic time or length scale. These complex systems are very common in nature (e.g., glassy systems, ecological systems, light from quasars) and they are characterized by a temporal effect known as $1/f^\beta$ noise, and by a spatial structure with scale-invariant or self-similar (fractal) properties (BAK and TANG and references therein, 1989).

Recently many authors focused their attention on observational evidence of earthquake scale-invariance properties, and introduced models and theories to analyze the complex dynamics of the lithosphere and to describe the space and time evolution of seismicity patterns (e.g., KAGAN, 1994; SORNETTE and SORNETTE, 1989; TURCOTTE, 1995). The understanding of the physical mechanism underlying

---

[1] Istituto di Metodologie Avanzate di Analisi Ambientale, CNR, C. da S. Loja, Zona Industriale, 85050 Tito (PZ), Italy.

[2] Dip. di Scienze Fisiche, Università Federico II, Pad. 20 Mostra d'Oltremare, 80125 Naples, Italy.

the presence of scaling properties in earthquake dynamics is a very fascinating research topic. It is widely assumed that the earthquake dynamics is due to a stick-slip process involving sliding of the earth's crust along faults. When slip occurs at some location, the strain energy is released, and the stress propagates in the vicinity of that position. It has been suggested that the stick-slip picture can be viewed as a branching mechanism (KAGAN and KNOPOFF, 1987).

In this work we explore in a real case the occurrence of particular scaling laws that are typical fingerprints of the extended dissipative systems. Thus we analyze the earthquake activity in a seismogenetic area of Southern Italy, and in particular we focus our attention on two subzones of the investigated area the Irpinia Zone and the Potentino Zone (zone A and zone B, respectively (Fig. 1), that present different spatial and temporal clustering characteristics probably correlated to their different geological settings. The main goal of this paper is to detect scale-invari-

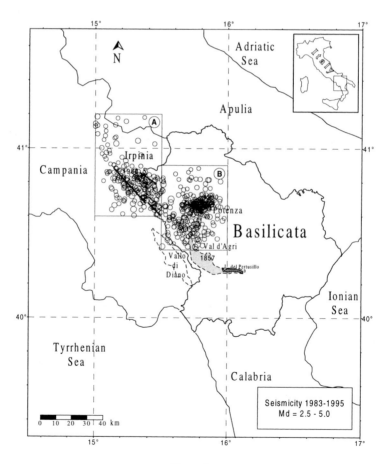

Figure 1
Spatial patterns of seismicity in two subzones of a seismogenetic area on the Southern Apennine chain.

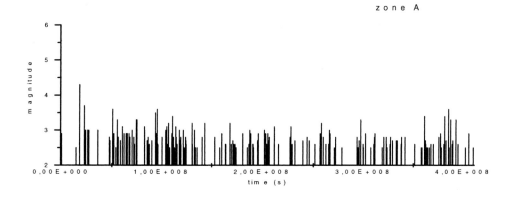

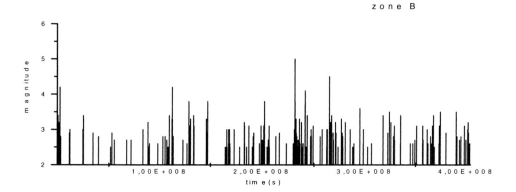

Figure 2
Frequency distribution of 475 earthquakes with magnitude $M \geq 2.5$ over the recorded time interval (Jan. 1, 1983 to Dec. 31, 1995). The data are extracted by the Catalogue of Istituto Nazionale di Geofisica and the two graphs are related to the seismic events which occurred in the two different subzones A and B depicted in Figure 1.

ance properties in the spatial and temporal seismicity patterns in a well-defined seismogenetic area and, as a consequence, to correlate the values of the key parameters in the revealed scaling laws with the main seismological settings of the active faults localized in the investigated area.

To this purpose a data set of earthquakes observed in the investigated area during the period 1983–1995 were extracted by the catalogue of the Istituto Nazionale di Geofisica. Because of the completeness of the catalogue which we tested by Gutenberg-Richter analysis, we consider only the distribution of the events with magnitude $M \geq 2.5$ (Fig. 2). Only one major event ($M_d = 5.0$ duration magnitude) is present in zone B; in zone A the seismicity is characterized by a low value of magnitude. The 1990 $M_d = 5.0$ earthquake in the Potentino generated a long sequence of aftershocks, as could be seen in the monthly cumulative number of events (Fig. 3). The Irpinia seems to have a constant rate of seismicity, compared

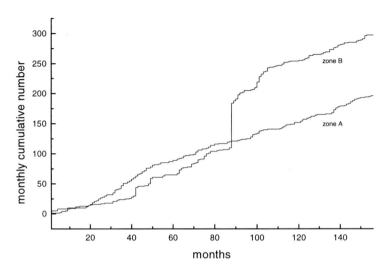

Figure 3
Monthly cumulative earthquake frequency distribution, for a magnitude threshold of 2.5. A spike is detected between 1990 and 1991 in the seismicity of Zone B when an increase of the seismicity related to the fault transverse oriented with respect to the Apenninic chain occurred.

with that of the Potentino. The Gutenberg-Richter analysis of the two earthquake sequences has given a b-value of 1.49 for zone A, and 1.08 for zone B (Fig. 4).

In each subzone of the seismogenetic area we characterize the $1/f^\beta$ temporal fluctuations of the occurrences of earthquakes by estimating the power spectrum density in its low-frequency range with counting statistics methodology (MAESMANN et al., 1993), and we also show the spatial self-similarity of the seismic activity with the calculation of the fractal dimension of the epicentre aggregate (VICSEK, 1992).

The paper is thusly organized: in section 2 the geological and seismological settings of the two seismic active areas of the Southern Apennine chain are outlined; in section 3 the mathematical background of the counting statistics approach is briefly described; in section 4 the procedures to detect scale-invariance laws in the spatial domain are presented, and finally, in section 5 our results and conclusions are discussed.

2. Geological and Seismological Settings

The Southern Apennine chain represents an African-verging fold and thrust belt build on during the Neogene tectogenesis. Starting in the middle Miocene and extending to the upper Pliocene, several compressional tectonic phases (PATACCA et al., 1988), associated with the collision between Africa and Europe, caused progres-

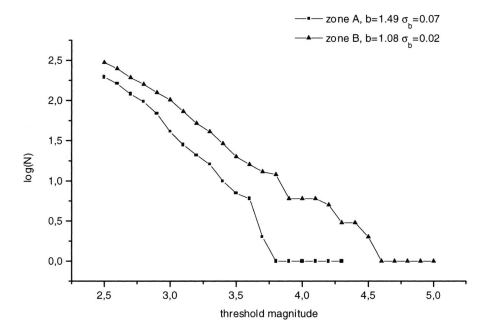

Figure 4
Gutenberg-Richter laws for the earthquake sequences of Figure 2.

sive thrusting and piling of different tectonic units corresponding to different paleogeographic domains, toward stable external domains of the Apulo-Adriatic foreland. Deformation front direction was NW-SE and migrated toward NE.

During the Quaternary, the Southern Apennines were affected by an important extensional tectonic phase, with a NE-SW extensional trend which caused further chain fragmenting into several isolated blocks. In the same period, counter-Apenninic regional transcurrent faults are generated.

The Southern Apennine chain is one of the most seismically active areas of the Mediterranean region. In particular, in this area major earthquakes are generally related to the great Apenninic normal faults, while moderate seismic events are generated by faults perpendicularly oriented with respect to the Apenninic chain (PANTOSTI and VALENSISE, 1990). Historical seismicity pattern confirms intense regional seismic activity and related complexity in crustal faulting. Most moderate and large earthquakes show a dip-slip mechanism, consistent with the actual extensional tectonics of the region (GASPARINI et al., 1985; WESTAWAY and JACKSON, 1987).

From a comparative analysis of both tectonic elements and historical and recent seismicity features, ALESSIO et al. (1995) identified in the Campania-Basilicata area at least four seismogenetic zones which exhibit activity with different features. In detail, they recognize the Irpinia, Potentino, Lagonegro and Vulture zones. In this work we are confined to analyzing the first two zones because the number of earthquakes in the last two areas is not sufficient to warrant statistical analysis (Fig. 1).

(A) *Irpinia Zone*

In this area (Fig. 1), three events with $M \approx 7.0$ (1694, 1930 and 1980) occurred in the last four centuries. In particular, the November 23, 1980, normal-faulting earthquake (surface-wave magnitude $M_s = 6.9$; seismic moment $M_0 = 26 \times 10^{18}$ Nm) was one of the largest events observed in the Southern Apennines in this century (WESTAWAY and JACKSON, 1987; BERNARD and ZOLLO, 1989; AMATO and SELVAGGI, 1993). It was the first well-documented example of surface faulting in the Italian peninsula. The seismological analysis revealed that the event was characterized by at least three different rupture episodes occurring at 0 s, ≈ 20 s and ≈ 40 s.

A particular seismogenetic feature of the Irpinia area is the substantial absence of events of moderate magnitude ($4.0 \leq M \leq 5.0$) (MULARGIA and TINTI, 1987; CONSOLE *et al.*, 1996). For this area, indeed, PANTOSTI and VALENSISE (1990) hypothesize the existence of a "characteristic earthquake" with $M \approx 7.0$ that releases completely the deformation energy of the area without producing a consistent energy release through significant aftershock sequences.

(B) *Potentino Zone*

In this area (Fig. 1), the seismic activity is related to a faulting system transverse to the great Apennine faults located on the northern side of the town of Potenza. This area is characterized by a local seismicity with low and medium magnitudes. In fact, these tectonic structures typically are responsible for nondestructive earthquakes.

In recent years this faulting system was responsible for two moderate seismic events which occurred on 5 May, 1990 ($M_d = 5.0$, ING-National Institute of Geophysics) and on 26 May, 1991 ($M_d = 4.5$) near Potenza. The temporal and spatial proximity of the 1990 and 1991 events to the great 1980 Irpinia-Basilicata earthquake adds great interest to our analysis. In particular the main features of the 5 May 1990 mainshock are unusual for the Apennines. These are characterized by a strike-slip mechanism and an estimate of scalar moment of $4-5 \times 10^{24}$ dyne-cm. The fault area outlined by the aftershocks extends approximately 20 km in length and 10 km in depth, making it significantly larger than that expected for a $M_d = 5.0$ earthquake. The aftershocks were concentrated between 15 and 25 km depth, which is deeper than the well-determined focal depth in the Central and Southern Apennines. Moreover, the 1990 and 1991 Potenza earthquakes may have had unusually low stress drops (EKSTRÖM, 1994).

3. Temporal Distribution of the Seismic Activity

In this section we apply an innovative technique to detect scale-invariance laws in the time dynamics of earthquake sequences. It can be demonstrated (BITTNER et al., 1996) that the temporal fluctuations of earthquake activity are characterized by a power spectrum density $S(f)$ decaying as $f^{-\beta}$ at low frequencies. The exponent β could be viewed as representative of the temporal clustering of the earthquake sequence, so that a β close to zero means that the time series is a realization of a point Poissonian process and could be considered purely random; adversely, a $\beta > 0$ provides information on the clustering feature of the time series. Thus the estimation of the spectral exponent is a tool used to discriminate random from clusterized earthquake sequences.

The rationale of the method consists in the relationship between the second-time derivative of the variance-time curve (VTC), defined as the variance of the counts in the time interval $\Delta t = (t_1, t_2)$, and the autocovariance of the point process, describing the earthquake sequence (BITTNER et al., 1996).

If a discriminating magnitude level M_0 has been introduced, the point process of the recorded earthquakes with $M(t_i) > M_0$ can be expressed by a finite sum of Dirac's delta functions centered on the occurrence times t_i.

$$y(t) = \sum_{i=1}^{n} \delta(t - t_i). \tag{1}$$

For the analysis $y(t)$ is assumed to be statistically stationary.

The actual number of events $N(\Delta t)$ occurring in a time interval $\Delta t = (t_1, t_2)$ is expressed as

$$N(\Delta t) = \int_{t_1}^{t_2} \sum_{j=1}^{n} \delta(t - t_j) \, dt. \tag{2}$$

The second time derivative of the variance of counts $\text{Var}(N(\Delta t))$, the so-called variance-time curve (VTC), is related to the autocovariance function of $y(t)$, $C_y(\Delta T)$ by the following equation

$$C_y(\Delta t) = \frac{1}{2} (\text{Var}[N(\Delta t)])'' \tag{3}$$

and the determination of the low frequency part of the spectrum $S_y(f)$ is related to the experimental determination of the $\text{Var}(N(\Delta T))$. If the VTC follows a power law within certain limits $f_{\min} < f < f_{\max}$ we have

$$\text{Var}[N(\Delta t)] \propto (\Delta t)^{1+\beta}, \quad \text{with } \beta < 1, \tag{4}$$

then it can be shown, using the Wiener-Chinchin theorem (MAESMANN et al., 1993), that the spectrum $S_y(f)$ scales within this range as

$$S_y(f) \propto f^{-\beta}. \tag{5}$$

The VTC is defined by the variance of counts for time interval lengths Δt as

$$\text{Var}[N(\Delta t)] = \langle N^2(\Delta t) \rangle - \langle N(\Delta t) \rangle^2$$

with $\langle \cdots \rangle$ denoting the expectation values. To estimate $\text{Var}(N(\Delta t))$, the entire observation time T is divided into k counting windows of duration Δt with $T = k \, \Delta t$, and the variance of counts is determined for this particular window Δt. This is repeated for different values of Δt. Plotting $\text{Var}(N(\Delta t))$ versus Δt on a log-log scale, it is possible to fit by linear regression the VTC and to deduce the exponent β.

The counting statistics method was used to determine the low frequency part of the spectrum $S(f)$ over the entire observation time T with interval lengths from 30 s to 10^6 s. The low-frequency part of the power spectrum density is determined by plotting the variance of the counts versus the interval length Δt on a log-log scale.

Performing this method on the earthquake sequence observed in the two zones, A and B, of the selected seismogenetic area, we observed different temporal clustering characteristics. Concerning zone A, fitting the data by linear regression (Fig. 5) we can conclude that at every interval length the earthquakes are randomly distributed, as a Poisson process, and the exponent $1 + \beta \approx 1$, with $\beta \approx 0$. Also, the VTC slope varies with the magnitude level very slowly (Fig. 6) with a mean value of 1.006. Thus, the seismic process in zone A could be viewed Poissonian at every length scale and at every magnitude threshold.

In contrast to zone A, two scaling regions can be detected for zone B (Fig. 5). Fitting the data by linear regression, we observe that at interval lengths less than the time threshold $\tau_c \approx 1500$ s the earthquakes are distributed as a Poisson process, with $\beta \approx 0$. Above this threshold, for medium range time intervals, the exponent $1 + \beta \approx 1.37$ (for $M_0 = 2.5$), with $\beta \approx 0.37$. The interval size τ_c is small compared with the average inter-event time ($\approx 1.4 \, M_s$), so that the events to find considerable variations of counts are sparse. Thus, in the scaling region the low-frequency part of the spectrum scales as $S(f) \propto f^{-0.37}$. In contrast to the zone A, the VTC slope varies with the magnitude level: higher threshold magnitudes M_0 correspond to lower exponent $1 + \beta$ in the scaling region, at higher time intervals the exponent is quite high but approaches unity with increasing magnitude level because of lower counts (Fig. 6).

This detected scaling law is a typical fingerprint of extended dissipative dynamical systems. It implies that the unpredictability of earthquakes observed in zone B is related to the infinite number of degrees of freedom rather than the high sensitivity to initial conditions of a few nonlinear deterministic equations (BAK and TANG, 1989).

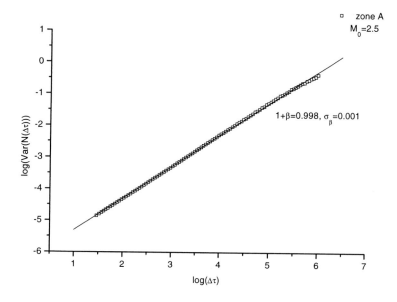

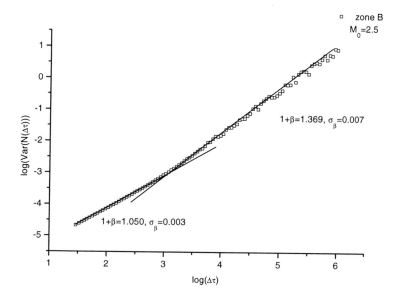

Figure 5
The variance of counts for the earthquake occurrences plotted on a decadic log-log scale versus the counting interval from 30 s to 10^6 s.

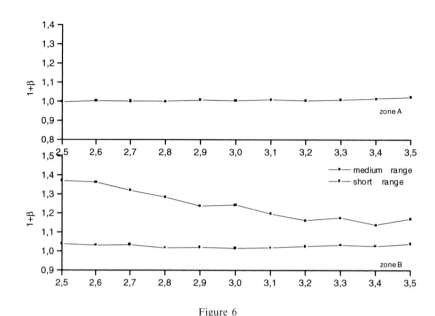

Figure 6
VTC slopes for different threshold magnitudes, calculated for short-range time interval (<1 hour) and medium (10^3 s to 10^6 s); the confidence intervals are not marked because the relative bars do not exceed the symbol size.

4. Fractal Analysis of Spatial Earthquake Distribution

Now we approach the problem of the spatial clustering of the earthquakes and we use methods to highlight self-similar behaviors. As in the previous section we analyze the seismicity in two different zones of a seismogenetic area of the Southern Apennine chain. The self-similar structure of seismic events has been under investigation in recent years (MEAKIN, 1991; TURCOTTE, 1995; HIRATA, 1989; XU and BURTON, 1995). In particular, the analysis of the spatial and temporal fractal dimensions of earthquake aggregates has been applied by many authors as a powerful tool to describe the main features of Italian's seismicity (GODANO and CARUSO, 1995; DE RUBEIS et al., 1993).

To analyze the self-similarity (scale-invariance) of the earthquake set, we estimated the fractal dimension D. Although the fractal dimension of a discrete set is zero, the strict mathematical definition in terms of its topology (Hausdorff-Besicovitch dimension) is not the one in common use in geophysics (MAIN, 1996). In this paper we started with the preliminary assumption that the statistical number size distribution for numerous objects can be considered fractal when the number of objects N with a characteristic linear dimension, r, satisfies the following relation:

$$N \propto r^{-D}$$

where D is the fractal dimension. Obviouisly this relation is only approximately applicable to experimental data sets (i.e., in our case, the frequency-size distribution of earthquakes) as discussed by TURCOTTE (1995).

With this assumption, we calculated the fractal dimension or the capacity dimension (D_B) for seismic events (MEAKIN, 1991; TURCOTTE, 1995; HIRATA, 1989) using the following procedure: to define it, we cover the aggregate with square elements, each of side L. Let $N(L)$ be the minimum number of square elements needed to cover the epicenter set. If the set is fractal, varying the length of L, we obtain the relation

$$N(L) = kL^{-D_B}.$$

To calculate D_B, we used the box-counting method. Finally we show the dependence of the fractal dimension upon the magnitude level; in every zone, greater magnitude level corresponds to lower fractal dimension (Fig. 7). Selecting a low threshold magnitude, in the fractal dimension calculation many events are involved, especially the aftershocks correlated with the strongest earthquakes. For this reason, the events appear more clusterized in space and the value of the fractal dimension is high. Increasing the threshold magnitude only a limited number of events with high magnitude is selected; these events present lower clustering of epicenters.

According to other authors (HIRATA, 1989; TURCOTTE, 1995) the analysis of the seismicity evolution and the search of possible precursors can be performed by studying the temporal variation of the fractal dimension, the b-value (Gutenberg-

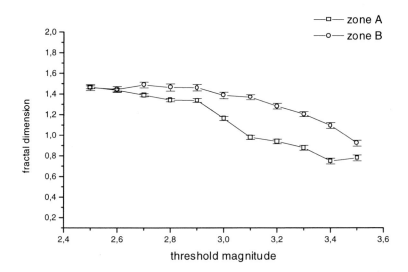

Figure 7
Fractal dimension of epicenter aggregates localized in the two subzones depicted in Figure 1. Only seismic events with magnitude $M \geq 2.5$ has been included in the statistical analysis.

Richter parameter) and other key parameters of earthquake dynamics. Thus our findings could be considered as a first preliminary step of a research activity devoted to study the fluctuations with time of the most important dynamical parameters of seismicity patterns observed on the Southern Apennine chain. Future studies could integrate the above results with information obtained from other kinds of research activities on short-term earthquake prediction carried out in Southern Italy (CUOMO et al., 1995, 1997).

5. Conclusions

The study of the spatial and temporal patterns of seismicity observed during the period 1983–1995 reveals the presence of two seismic zones with different scale invariance properties, probably due to their different geological and seismological settings. The main features of the dynamic system under study are easily depicted using the key parameters of more recurrent scaling laws observed in the spatial and temporal domains. The Irpinia zone has been characterized by a $\beta = 0$ at every scale time length, so that the distribution of earthquakes in that zone is Poissonian. Adversely, the Potentino zone has been characterized by a power spectrum, scaling at low frequencies as $1/f^\beta$, with $\beta > 0$. Furthermore, there exists a critical time length (≈ 1500 s) under which the spectrum could be viewed as flat, with $\beta = 0$, and the geophysical process underlying the seismic phenomena could be considered Poissonian. In the spatial domain the estimate of the fractal dimension highlights the self-similar structure of seismic events in both subzones of seismogenetic area. The temporal scale invariance law observed in the earthquake sequences of zone B and the spatial scale invariance properties revealed in both the zones are typical fingerprints of the extended dissipative system and, as a consequence, the predictability of the seismotectonic environment of the Southern Apennine is strongly limited by the large number of degrees of freedom. Future studies devoted to analyze the changes throughout the time of these dynamical parameters of seismicity could possibly contribute to earthquake prediction research.

REFERENCES

ALESSIO, G., ESPOSITO, F., GORINI, A., and PORFIDO, S. (1995), *Detailed Study of the Potentino Seismic Zone in the Southern Apennines*, Tectonophysics 250, 113–134.

AMATO, A., and SELVAGGI, G. (1993), *Aftershock Location and P-velocity Structure in the Epicentral Region of the 1980 Irpinia Earthquake*, Annali di Geofisica XXXVI, 1, 3–15.

BAK, P., and TANG, C. (1989), *Earthquakes as a Self-organized Critical Phenomenon*, J. Geophys. Res. 94, B11, 15,635–15,637.

BERNARD, P., and ZOLLO, A. (1989), *The Irpinia 1980 Earthquake: Detailed Analysis of a Complex Normal Faulting*, J. Geophys. Res. 94, 1631–1647.

BITTNER, H. R., TOSI, P., BRAUN, C., MAEESMAN, M., and KNIFFKI, K. D. (1996), *Counting Statistics of f^{-b} Fluctuations: A New Method for Analysis of Earthquake Data*, Geol. Rundsch. *85*, 110–115.

CONSOLE, R., MONTUORI, C., and MURRU, M. (1996), *Seismicity Rate Change before the Irpinia, November 23, 1980 Earthquake*, Proceeding of European Seismological Commission, Reykjavik, Iceland.

CUOMO, V., DI BELLO, G., LAPENNA, V., MACCHIATO, M., and SERIO, C. (1995), *Parametric Time Series Analysis of Extreme Events in Earthquake Electrical Precursors*, Tectonophysics *262*, 159–172.

CUOMO, V., LAPENNA, V., MACCHIATO, M., and SERIO, C. (1997), *Autoregressive Models as a Tool to Discriminate Chaos from Randomness in Geoelectrical Time Series: An Application to Earthquake Prediction*, Annali di Geofisica *XL*, 2, 385–400.

DE RUBEIS, V., DIMITRIU, P., PAPADIMITRIOU, E., and TOSI, P. (1993), *Recurrent Patterns in the Spatial Behaviour of the Italian Seismicity Revealed by the Fractal Approach*, Geophys. Res. Lett. *20*, 1911–1914.

EKSTRÖM, G. (1994), *Teleseismic Analysis of the 1990 and 1991 Earthquakes near Potenza*, Annali di Geofisica *XXXVII*, 6, 1591–1599.

GASPARINI, C., INNACCONE, G., and SCARPA, R. (1985), *Fault-plane Solution Seismicity of the Italian Peninsula*, Tectonophysics *117*, 59–78.

GODANO, C., and CARUSO, V. (1995), *Multifractal Analysis of Earthquake Catalogues*, Geophys. J. Int. *121*, 385–392.

HIRATA, T. (1989), *A Correlation between the b Value and the Fractal Dimension of Earthquakes*, J. Geophys. Res. *94*, 7507–7514.

KAGAN, Y. Y. (1994), *Observation Evidence for Earthquakes as a Nonlinear Dynamics Process*, Physica D *77*, 160–192.

KAGAN, Y. Y., and KNOPOFF, L. (1987), *Statistical Short-term Earthquake Prediction*, Science *236*, 1563.

MAESMANN, M., BOESE, J., CHIALVO, R. R., KOWALLIK, P., PETERS, W., GRUNEIS, F., and KNIFFKI, K. (1993), *Demonstration of $1/f^\beta$ Fluctuations and White Noise in the Human Heart Rate by the Variance-time Curve: Implications for Self-similarity*, Fractals *1*, 312–320.

MAIN, I. (1996), *Statistical Physics, Seismogenesis, and Seismic Hazard*, Rev. Geophys. *34*, 433–462.

MEAKIN, P. (1991), *Fractal Aggregates in Geophysics*, Rev. Geophys. *29*, 317.

MULARGIA, F., and TINTI, S. (1987), *Seismic Sample Areas Defined from Incomplete Catalogues: An Application to the Italian Territory*, Phys. Earth Planetary Inst. *40*, 273–300.

PANTOSTI, D., and VALENSISE, G. (1990), *Faulting Mechanism and Complexity of the November 23, 1980, Campania Lucania Earthquake, Inferred from Surface Observations*, J. Geophys. Res. *95*, B10, 15, 329–15,341.

PATACCA, E., SCANDONE, P., BELLATALLA, M., PERILLI, N., and SANTINI, U. (1988), *L'Appennino Meridionale: mondello strutturale e palinspastica dei domini esterni.* Invited paper at the 74° Congresso Soc. Geol. It., 67–69.

SORNETTE, A., and SORNETTE, D. (1989), *Self-organizing Criticality and Earthquakes*, Europhys. Lett. *9* (2), 197.

TURCOTTE, D. L., *Fractals and Chaos in Geology and Geophysics* (Cambridge University Press, Cambridge 1995).

VICSEK, T., *Fractal Growth Phenomena* (World Scientific, Singapore 1992).

WESTAWAY, R., and JACKSON, J. A. (1987), *The Earthquake of 1980 November 23 in Campania-Basilicata (Southern Italy)*, Geophys. J. R. Astron. Soc. *90*, 375–443.

XU, Y., and BURTON, P. W. (1995), *Temporal Scaling Regions and Capacity Dimensions for Microearthquake in Greece*, Proceedings of Fractal in Natural Sciences, Toulouse.

(Received April 7, 1998, revised December 1, 1998, accepted December 12, 1998)

Variation of Permeability with Porosity in Sandstone Diagenesis Interpreted with a Fractal Pore Space Model

HANSGEORG PAPE,[1] CHRISTOPH CLAUSER[1] and JOACHIM IFFLAND[2]

Abstract—Permeability is one of the key rock properties for the management of hydrocarbon and geothermal reservoirs as well as for aquifers. The fundamental equation for estimating permeability is the Kozeny-Carman equation. It is based on a capillary bundle model and relates permeability to porosity, tortuosity and an effective hydraulic pore radius which is defined by this equation. Whereas in clean sands the effective pore radius can be replaced by the specific surface or by the grain radius in a simple way, the resulting equations for permeability cannot be applied to consolidated rocks. Based on a fractal model for porous media, equations were therefore developed which adjust the measure of the specific surface and of the grain radius to the resolution length appropriate for the hydraulic process. These equations are calibrated by a large data set for permeability, formation factor, and porosity determined on sedimentary rocks. This fractal model yields tortuosity and effective pore radius as functions of porosity as well as a general permeability-porosity relationship, the coefficients of which are characteristic for different rock types. It can be applied to interpret the diagenetic evolution of the pore space of sedimentary rocks due to mechanical and chemical compaction with respect to porosity and permeability.

Key words: Permeability, sandstone, fractals, diagenesis.

Introduction

Understanding fluid flow, heat and mass transport, and geochemical fluid-rock interaction is important in reservoir simulation studies of groundwater, hydrocarbons and geothermal energy. This requires information regarding several petrophysical parameters, above all porosity ϕ and permeability k. Permeability varies over nine decades for different kinds of sandstone. Usually minimal information is available on *in situ* permeability. It is therefore attractive to infer permeability from properties which are usually better known, such as porosity, specific surface or grain radius. PAPE *et al.* (1999) discuss the various theoretical and empirical approaches that have been used for this purpose. If information is

---

[1] Geoscientific Research Institute (GGA) of the German Geological Surveys, Stilleweg 2, D-30655, Hannover, Germany, e-mail: c.clauser@gga-hannover.de
[2] LUNG Mecklenburg-Vorpommern, Pampower Str. 66/68, D-19061 Schwerin, Germany, e-mail: lung-sn@um.mv-regierung.de

available on the distribution of the radii of pore throats, network or percolation theory can be applied (DAVID et al., 1990) to infer permeability within a factor of two or better. The fractal approach presented here can be used as an alternative to these methods. It can be applied in the typical situation, i.e., with much less available information. In a comprehensive paper PAPE et al. (1999) discuss how the fractal concept can be employed to calculate permeability from a combination of different borehole measurements.

In this paper we start from well-known equations for permeability which are based on a rock model with smooth pores and grains. These equations are power laws with integer exponents. Next, an alternative model is applied which is characterized by a fractal interface between grains and pores. This approach yields modified equations with broken exponents. Both types of relationships are examined using a large petrophysical data set obtained for sandstones. It will be shown that the fractal approach yields a better agreement between calculated and measured permeability than the simple model.

The fundamental Kozeny-Carman equation (KOZENY, 1927; CARMAN, 1937, 1948, 1956) relates permeability k to porosity ϕ and tortuosity T as well as to the effective hydraulic pore radius r_{eff} of the capillary system:

$$k = (\phi/T)(r_{\text{eff}})^2/8. \tag{1}$$

In this equation it is assumed that the pores are bundles of smooth, cylindrical but tortuous capillaries of radius r_{eff} (Fig. 1a). This mean model radius cannot be determined directly. However it can be substituted either by S_{por}, the specific surface normalized by the pore volume,

$$r_{\text{eff}} = 2/S_{\text{por}}, \tag{2}$$

or by the grain radius r_{grain} and the porosity ϕ,

$$r_{\text{eff}} = (2/3)r_{\text{grain}}\phi/(1-\phi). \tag{3}$$

In equation (2) the effective pore radius is calculated as twice the ratio of the model capillary's volume and its surface. In analogy, equation (3) follows for a model porous medium made up of smooth spherical grains (Fig. 1b), if the ratio of the sphere's volume to its surface, weighted by the void ratio $\phi/(1-\phi)$, is inserted into equation (2) for $1/S_{\text{por}}$.

The ratio T/ϕ is the formation factor F and can be determined from electrical measurements. The reciprocal of the formation factor describes the effective porosity with respect to transport processes such as fluid flow, electrical conductance, and diffusion. Numerical simulations on two-dimensional networks by DAVID (1993) showed that the so-called 'network tortuosity' for hydraulic flow was $1\frac{1}{2}$ times larger than the 'network tortuosity' for electrical current. While this should be kept in mind, it poses no real problems in fractal model theory since the equations are mainly governed by the exponents. The coefficients are determined by calibration with data.

If F is not available, a simple relationship between formation factor and porosity can be applied, the so-called Archie's first law (ARCHIE, 1942):

$$F = A\phi^{-m}, \qquad (4)$$

with $A = 0.7$ and $m = 2$ for an average sandstone.

Inserting equations (2) and (3) into equation (1) yields the following modified Kozeny-Carman equations:

$$k = (1/2)(\phi/T)(S_{por})^{-2}, \qquad (5)$$

and

$$k = (1/18)T^{-1}(r_{grain})^2 \phi^3/(1-\phi)^2 \qquad (6)$$

However, equations (5) and (6) do not yield a satisfactory fit to data of natural sandstones, except for very clean types with a large porosity. If the grains are overgrown by secondary minerals during diagenesis, r_{eff} cannot be substituted by equations (2) and (3) because the measures of the specific surface, either measured directly or determined from the effective pore radius or the grain radius, depend on

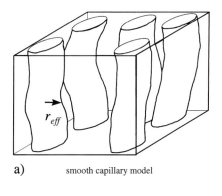

a) smooth capillary model

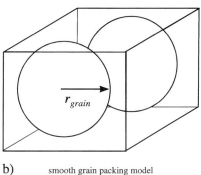

b) smooth grain packing model

Figure 1
Simple models for porous media composed of: (a) smooth capillaries and (b) smooth spheres.

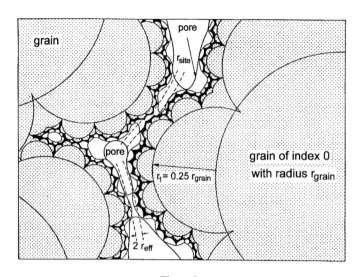

Figure 2
Cartoon showing a sedimentary rock according to the "pigeon hole" model (PAPE et al., 1987a) composed of geometrical pores with radius r_{site} and hydraulic capillaries with effective radius r_{eff}.

the resolution power of the physical processes used for measuring. The real specific surface and the model area corresponding to spheres of radius r_{grain} are measured by a method with higher and lower resolution, respectively, compared to the resolution of the flow process, which is characterized by a resolution length equal to r_{eff}. Therefore, the measured values of S_{por} and specific model surface corresponding to spheres of radius r_{grain} must be converted before they can be inserted into equations (2) and (3).

Model Theory

The pore space of natural rocks is hierarchically structured over a wide range of scales. This is reflected by a special fractal pore space model, the so-called "pigeon hole" model (PAPE et al., 1982 PAPE et al., 1987a) (Fig. 2). Based on this model, effective pore radius and permeability can be calculated assuming a multi-fractal structure with the predominant fractal dimension $D = 2.36$. As the capillaries have constrictions, it is necessary to distinguish between hydraulic radii r_{eff} and geometric pores r_{site} (Fig. 2). The geometrical pore sites are connected by hydraulic capillary channels of radius r_{eff}. The three radii r_{grain}, r_{site} and r_{eff} are the size parameters of the porous medium. They are related by an empirical expression, which is valid for a great variety of sandstones (PAPE et al., 1984):

$$r_{\text{grain}}/r_{\text{site}} = (r_{\text{grain}}/r_{\text{eff}})^{c_1}, \qquad (7)$$

where $c_1 = 0.39$ if $r_{\text{grain}}/r_{\text{eff}} > 30$. Our analysis of the large data set of sandstone samples confirms this mean value for c_1 and also yields the general expression $c_1 = 0.263\phi^{-0.2}$ for $r_{grain}/r_{eff} > 30$.

A measure $M(\lambda_1)$, obtained for the resolution length λ_1, is converted to a measure $M(\lambda_2)$ for a different resolution length λ_2 by:

$$M(\lambda_1)/M(\lambda_2) = (\lambda_1/\lambda_2)^{(D_t - D)}, \qquad (8)$$

where D_t is the topological dimension of the measure.

For instance, the dependence of the specific surface normalized to total pore volume S_{por} on the resolution power is given by:

$$S_{por}(\lambda_1)/S_{por}(\lambda_2) = (\lambda_1/\lambda_2)^{(D_t - D)}. \qquad (9)$$

$D_t = 2$ is the topological dimension of a surface.

In petrophysical research, the specific surface S_{por}^{BET} is measured by the Brunauer-Emmett-Teller (BET) method of nitrogen adsorption (BRUNAUER et al., 1938). Thus, the resolution length is of the order of the size $d(N_2)$ of nitrogen molecules. The specific surface in equation (2), which is "seen" by the hydraulic process, has a resolution length of the order of r_{eff} and can be written as $S_{por}(r_{eff})$, as explained below. Substitution of $S_{por}(r_{eff})$, S_{por}^{BET}, $d(N_2)$, and r_{eff} for $S_{por}(\lambda_1)$, $S_{por}(\lambda_2)$, λ_1, and λ_2, respectively, and replacing $d(N_2)^{(D_t - D)}$ by $1/const1$ yields the relationship:

$$S_{por}(r_{eff}) = const1\, S_{por}^{BET} r_{eff}^{(2-D)}, \qquad (10)$$

with $D = 2.36$ and $const1 = 0.141$, determined from calibration with laboratory measurements (PAPE et al., 1987a).

Inserting equation (10) into equation (2) yields

$$r_{eff} = (14/S_{por}^{BET})^{1/(3-D)}. \qquad (11)$$

Equation (11) relates the effective pore radius to the specific surface for porous media with a fractal structure, in contrast to equation (2) which was derived for the capillary model with smooth surfaces.

Next we will show how the effective pore radius can be estimated on the base of data on grain radius and porosity. We present two derivations in which either porosity ϕ or the length L of the hydraulic flow path is regarded as a fractal. The first derivation is based on an empirical expression (PAPE et al., 1987a) which relates porosity to r_{site} and r_{grain}:

$$\phi = 0.5(r_{grain}/r_{site})^{(3-D)}. \qquad (12)$$

Equation (12) describes how ϕ decreases with decreasing r_{site} during the diagenetic compaction of sandstones (PAPE et al., 1984). Inserting equation (7) relating r_{eff}, r_{site}, and r_{grain}, into equation (12), yields the desired expression for r_{eff}:

$$r_{eff} = r_{grain}(2\phi)^{1/(c_1(3-D))}, \qquad (13)$$

with $c_1 = 0.263\,\phi^{-0.2}$ for $r_{grain}/r_{eff} > 30$, as before.

An alternative expression is useful if tortuosity T is available from electrical measurements. The basic assumption is that the measure of the surface of the grains equals the measure of the surface of the pore sites, provided that both are measured with a low resolution length. For cylindrical pores with radius r_{site}, the specific surface $S_{por}(r_{site})$, measured with the resolution length r_{site}, equals $2/r_{site}$ as in equation (2). The specific surface $S_{por}(r_{grain})$, measured with the resolution length r_{grain}, can be expressed by the specific surface $S_{solid}(r_{grain})$, which is normalized by the volume of the solids and which equals $3/r_{grain}$ for spherical grains with radius r_{grain}:

$$S_{por}(r_{grain}) = S_{solid}(r_{grain})(1-\phi)/\phi = (3/r_{grain})(1-\phi)/\phi. \tag{14}$$

Replacing $S_{por}(r_{grain})$ by $S_{por}(r_{site}) = 2/r_{site}$ yields

$$r_{site} = (2/3)r_{grain}\phi/(1-\phi). \tag{15}$$

The effective pore radius r_{eff} is always smaller than r_{site}. This is expressed by the so-called constrictivity $\sigma < 1$ (VAN BRAKEL, 1975; PAPE et al., 1987a) which accounts for the fact that only a certain fraction of the pore volume is involved in the flow process. While constrictivity cannot be measured directly, it can be inferred from tortuosity T if the latter is assumed to be composed of a constrictivity term σ^{-1} and a term $\Lambda^2 > 1$ which accounts both for the relative increase in hydraulic flow path length and for the associated increase in volume of the tortuous hydraulic capillary:

$$T = \Lambda^2 \sigma^{-1}. \tag{16}$$

(See the Appendix for a discussion of whether Λ or Λ^2 should be used in the definition of tortuosity.) Thus, in order to approximate r_{eff}, porosity in the nominator of equation (15) is substituted by an "effective porosity" equal to ϕ/σ^{-1}:

$$r_{eff} = (2/3)r_{grain}\phi\sigma/(1-\phi) = (2/3)r_{grain}\phi\Lambda^2 T^{-1}/(1-\phi). \tag{17}$$

The flow path length increase Λ is inferred from the fractal scaling law (equation 8) where $D_t = 1$ and D is substituted by $D_L = D' - 1$ where D_L is the fractal dimension of the hydraulic flow path and D' is the fractal dimension of the surface of the hydraulic capillary. If $L(\lambda_2)$ is the length of the straight pore with resolution $\lambda_2 = r_{grain}$, and $L(\lambda_1)$ is the length of the real flow path expressed with the proper resolution length $\lambda_1 = r_{eff}$, Λ can be expressed by:

$$\Lambda = L(\lambda_1)/L(\lambda_2) = (\lambda_1/\lambda_2)^{(1-D_L)} = (r_{eff}/r_{grain})^{(2-D')}. \tag{18}$$

The shape of this surface is smoother than the surface of the real pore space. Therefore D' is smaller than the general fractal dimension D of the porous rock. In order to account for this, D' will be approximated by an expression which places the roughness of the surface of the hydraulic capillary midway between a perfectly smooth surface and the roughness of the real pore walls:

$$2 - D' = (2 - D)/2. \tag{19}$$

Inserting equation (19) into equation (18) yields:

$$\Lambda^2 = (r_{\text{eff}}/r_{\text{grain}})^{2(2-D')} = (r_{\text{eff}}/r_{\text{grain}})^{(2-D)}. \tag{20}$$

Inserting this expression for Λ^2 into equation (17) yields the desired second expression for r_{eff}:

$$r_{\text{eff}} = r_{\text{grain}}((2/3)T^{-1}\phi/(1-\phi))^{1/(D-1)}. \tag{21}$$

The tortuosity T in equation (21) can be determined from measurements of the electrical resistivity. Alternatively it also can be calculated from porosity and Archie's law, as will be shown later. Equations (13) and (21), relating r_{eff} to r_{grain} and ϕ for rough pore walls, are the fractal analogues of equation (3) which had been derived for a model pore space made up of smooth spheres.

Calibration and Validation of the Fractal Effective Pore Radius Relationships with Petrophysical Measurements in Sandstones

On the one hand, equations (11), (13), and (21) for the effective pore radius are based on fractal theory. On the other hand they also contain empirical relationships. Therefore they must be calibrated with laboratory measurements on core samples.

Permeability, porosity, and formation factor were measured on samples of clean and shaly sandstones from hydrocarbon reservoirs and aquifers in northwest Germany (DEBSCHÜTZ, 1995; KULENKAMPFF, 1994 and additional personal communications from the Petrophysical Research Group at the Institute of Geophysics, University of Clausthal, Germany). A comparison between the value for r_{eff} obtained by inserting measured permeability into the Kozeny-Carman equation (equation 1) and the different theoretical predictions is shown in Figure 3 for the expressions based on specific surface (equations 2 and 11), and in Figure 4 for the expressions based on porosity (equations 3, 13, and 21). Figure 3 shows that the fractal relationship (equation 11) is superior to the smooth-capillary model (equation 2). The linear regressions corresponding to the two expressions converge for small pore radii only. This is reasonable, since the appropriate resolution length for measuring the specific surface, i.e., $\lambda = r_{\text{eff}}$, approaches the small resolution length of the nitrogen adsorption method for small values of r_{eff}. Figure 4 shows that the two fractal relationships (equations 13 and 21) are superior to the smooth grain packing model (equation 3). The linear regressions corresponding to the three expressions converge for large pore radii only. This is again reasonable since the resolution length of the grain radii approaches the resolution length of the pore radii for large pore radii.

Figure 5 is a cross-plot of pore radii versus porosity for four sets of pore radii obtained by inserting measured permeability into equation (1) and from the expressions for the smooth spherical model (equation 3) and the two fractal expressions (equations 13 and 21). It demonstrates that the two fractal expressions yield good results over the entire range of porosity in the data, in contrast to the equation based on the smooth grain packing model.

Permeability Calculated by the Modified Kozeny-Carman Equations Based on the Fractal Concept

The fractal expressions for the effective pore radius can be inserted into the Kozeny-Carman equation (equation 1). If data on a specific surface is available, the relationship between pore radius and a specific surface (equation 11) can be combined with equation (1). The result is the so-called PARIS-equation (PAPE, RIEPE and SCHOPPER, 1982):

$$k = 475.3(\phi/T) S_{\text{por}}^{(-2/(3-D))} \cdot (\text{in } \mu\text{m}^2). \tag{22}$$

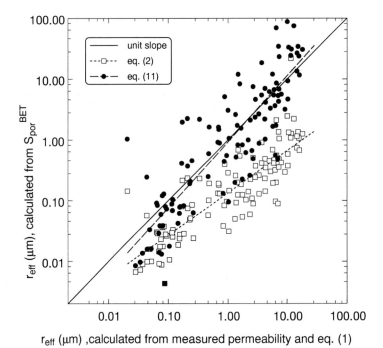

Figure 3
Comparison of different relationships between effective pore radius r_{eff} and specific surface S_{por} based on a simple capillary model (boxes) and a fractal model (circles). The straight line with unit slope indicates a one-to-one relationship.

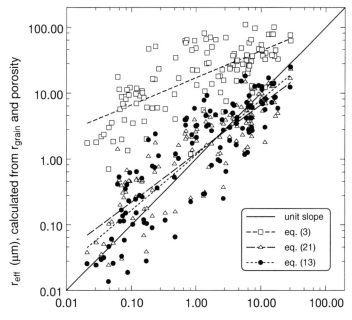

Figure 4
Comparison of three different relationships between effective pore radius r_{eff}, grain radius r_{grain}, and porosity based on a simple capillary model (boxes) and on two fractal models (triangles and circles). The straight line with unit slope indicates a one-to-one relationship.

If porosity and grain radius data are available, the relationships between pore radius and porosity as well as grain radius (equations 13 or 21) can be inserted into equation (1), yielding

$$k = (1/8)(\phi/T)(r_{grain})^2(2\phi)^{2/(c_1(3-D))}, \qquad (23)$$

with $c_1 = 0.263 \, \phi^{-0.2}$ for $r_{grain}/r_{eff} > 30$, as before, or

$$k = (1/8)(\phi/T)(r_{grain})^2((2/3)T^{-1}\phi/(1-\phi))^{2/(D-1)}. \qquad (24)$$

If equations (23) and (24) are applied to obtain permeability from borehole measurements, information on porosity, grain radius, and tortuosity is required. While porosity and tortuosity can be determined directly from various geophysical logs, information on grain radius is more difficult to obtain. Often it must be deduced from rock type. Tortuosity T can be calculated from ϕ and Archie's first law:

$$T/\phi = A\phi^{-m} \qquad (25)$$

using the following expressions for m and A from fractal model theory (PAPE and SCHOPPER, 1988):

$$-m = (0.67(D-2))/(c_1(D-3)) - 1, \quad \text{with} \quad 0.39 < c_1 < 1, \tag{26a}$$

$$A = 1.34/0.534^{-m+1}. \tag{26b}$$

Finally, PAPE *et al.* (1999) derived a general permeability-porosity relationship whose coefficients a, b, and c were calibrated with values of ϕ and k measured on different types of sedimentary, igneous, and metamorphic rocks:

$$k = a\phi + b\phi^{\exp 1} + c(10\phi)^{\exp 2}, \tag{27}$$

with $\exp 1 = m$, $\exp 2 = m + 2/(c_1(3-D))$, and $0.39 < c_1 < 1$.

The third term of this equation corresponds to equation (23). It is valid for porosities larger than 0.1; for smaller porosities the measured permeabilities exceed those predicted by equation (23). The reason for this is twofold: (1) the effective pore radii of the samples studied by PAPE *et al.* (1987a,b) do not decrease as rapidly with decreasing porosity as suggested by equation (13) for porosities lower than 0.1; (2) tortuosity tends towards a finite limiting value $T_{\max} = 10$ for porosities lower than 0.01. The first two terms in equation (27) account for this and expand the validity of equation (23) in the low porosity range. A complete derivation and justification of this approach is discussed in detail in PAPE *et al.* (1999).

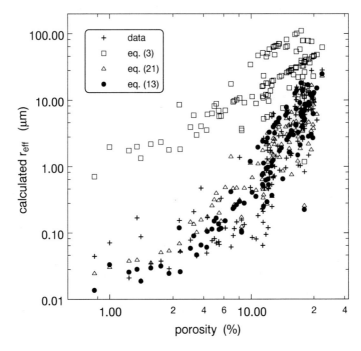

Figure 5

Effective pore radius r_{eff} plotted versus porosity for three different relationships between effective pore radius r_{eff}, grain radius r_{grain}, and porosity based on a simple capillary model (boxes) and two fractal expressions (triangles and circles). For comparison, crosses show data determined by inserting measured permeability into the Kozeny-Carman equation.

Figure 6 shows a comparison between permeability and porosity measured on different rock types with several curves calibrated from equation (27) with independent, large data sets for rocks that vary from clean and shaly sandstones to pure shales. Four curves correspond to the fractal model and have a shape defined by equation (27) with exponents according to a fractal dimension of $D = 2.36$. They are offset due to the different effective grain radii r_{grain} which determine the coefficients of the three terms. The smallest grain radius corresponds to the lowest and rightmost curve for pure shales.

In contrast, the data of BOURBIE and ZINSZNER (1985) from Fontainebleau sandstone define quite a different curve. This rock is characterized by strong quartz cementation with smooth surfaces corresponding to a fractal dimension of $D = 2$. Thus it is well suited for the smooth grain packing or capillary models (Fig. 1) which predict that permeability varies with porosity cubed (equation 6). Specifically, for porosities larger than 0.08, the permeability of Fontainebleau sandstone varies with porosity taken to a power of 3.05. For smaller porosities, however, the permeability of Fontainebleau sandstone varies with porosity taken to a power of 7.33. This corresponds to rough surfaces with a fractal dimension $D > 2$.

The permeability-porosity curves in Figure 6 reflect different diagenetic processes which tend to reduce porosity and permeability. We believe that the steep branch of the curves with an exponent of about 10 results from compaction during burial due to increasing overburden pressure. On the other hand, the smaller exponents (as in the large-porosity branch of the curve for Fontainebleau sandstone) may be due to cementation.

The variation of permeability during compaction was studied by DAVID et al. (1994) over a wide range of compaction pressure from 3 MPa–550 MPa. In most sandstones, except for Fontainebleau sandstone, there is a sharp inflection in the curves of permeability or porosity versus pressure. This inflection point marks the critical pressure where the crushing of grains and the collapse of pores begins. Below this critical pressure, grain dislocation is the dominant mechanism of compaction. Figure 7 plots the permeability and porosity data of these compaction experiments into the $k - \phi$ diagram of Figure 6. For different sandstones, open and full symbols mark the data below and above the critical pressure, respectively. The general permeability-porosity relationship below and above the critical pressure can be fitted by an exponential expression with an exponent equal to or larger than 10. Inspection of the permeability curves in closer detail reveals that their slope is not constant. In the context of our theory this implies that the exponent c_1 in equation (7) which relates r_{eff}, r_{site}, and r_{grain} is no longer a constant or a smooth function of porosity. This means that size parameters of the pore space (r_{site} and r_{eff}) decrease differently during different stages of the compaction: c_1 increases if r_{site} decreases more than r_{eff} otherwise c_1 decreases. Additionally, the average slope of the compaction data in the $k - \phi$ diagram is larger than predicted by calibrated versions of equation (27). This is due to the fact that r_{grain} does not remain constant during the compaction experiments because of grain crushing. In summary, a power law of the type of equation (27) is generally suited to interpret these compaction

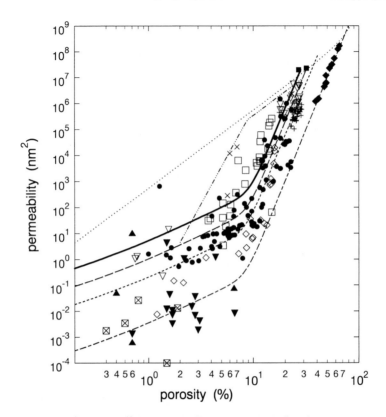

- — — — $k = 31\phi + 7\,463\phi^2 + 191(10\phi)^{10}$ fractal model for an average type of sandstone
- ———— $k = 155\phi + 37\,315\phi^2 + 630(10\phi)^{10}$ fractal model for Rotliegend sandstone northeast Germany
- ‑ ‑ ‑ ‑ ‑ $k = 6.2\phi + 1493\phi^2 + 58(10\phi)^{10}$ fractal model for a shaly sandstone
- — - — $k = 0.1\phi + 26\phi^2 + (10\phi)^{10}$ fractal model for a shale
- — · · — · · $k = 303(100\phi)^{3.05}$ for $\phi > 0.08$ Fontainebleau sandstone (BOURBIE and ZINSZNER, 1985)
  $k = 0.0275(100\phi)^{7.33}$ for $\phi \leq 0.08$ Fontainebleau sandstone (BOURBIE and ZINSZNER, 1985)
- · · · · · · · · $k = 0.5\,(r_{grain})^2 \phi^3/(1-\phi)^2$ smooth capillary model
- ■ sand (SCHOPPER, 1967)
- ◆ kaolinite (MICHAELS and LIN, 1954)
- × Fontainebleau sandstone
- ▽ Dogger sandstone
- □ Keuper sandstone
- + Bunter sandstone
- ● Rotliegend sandstone, northwest Germany
- ◇ Carboniferous sandstone
- ▼ Jurassic shale (SCHLÖMER and KROOSS, 1997)
- ▲ Rotliegend shale (SCHLÖMER and KROOSS, 1997)
- ⊠ Carboniferous shale (SCHLÖMER and KROOSS, 1997)

Figure 6
Permeability versus porosity for different consolidated and unconsolidated clean and shaly sandstones. Symbols indicate permeability and porosity measured on small sets of samples; the first four curves shown in the legend were calibrated to large sets of data (PAPE et al., 1999); the next two curves shown in the legend correspond to the model of BOURBIE and ZINSZNER (1985) and the smooth capillary model. Data not referenced individually are taken from DEBSCHÜTZ (1995), IFFLAND and VOIGT (1996), and KULENKAMPFF (1994).

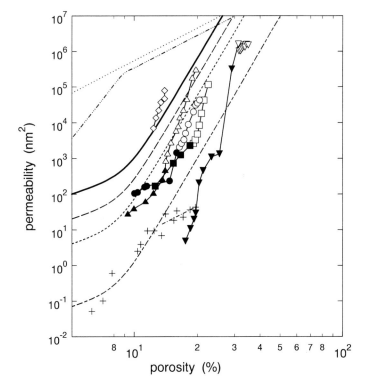

- - - - $k = 31\phi + 7\,463\phi^2 + 191(10\phi)^{10}$ fractal model for an average type of sandstone
——— $k = 155\phi + 37\,315\phi^2 + 630(10\phi)^{10}$ fractal model for Rotliegend sandstone northeast Germany
- - - - - - $k = 6.2\phi + 1493\phi^2 + 58(10\phi)^{10}$ fractal model for a shaly sandstone
– – – – $k = 0.1\phi + 26\phi^2 + (10\phi)^{10}$ fractal model for a shale
– · – ·· – $k = 303(100\phi)^{3.05}$ for $\phi > 0.08$ Fontainebleau sandstone (BOURBIE and ZINSZNER, 1985)
$k = 0.0275(100\phi)^{7.33}$ for $\phi \leq 0.08$ Fontainebleau sandstone (BOURBIE and ZINSZNER, 1985)
· · · · · · · · · $k = 0.5\,(r_{grain})^2 \phi^3/(1-\phi)^2$ smooth capillary model
△   Rothbach sandstone below critical pressure for grain crushing (DAVID et al., 1994)
▲   Rothbach sandstone above critical pressure for grain crushing (DAVID et al., 1994)
○   Berea sandstone below critical pressure for grain crushing (DAVID et al., 1994)
●   Berea sandstone above critical pressure for grain crushing (DAVID et al., 1994)
□   Adamswiller sandstone below critical pressure for grain crushing (DAVID et al., 1994)
■   Adamswiller sandstone above critical pressure for grain crushing (DAVID et al., 1994)
▽   Boise sandstone below critical pressure for grain crushing (DAVID et al., 1994)
▼   Boise sandstone above critical pressure for grain crushing (DAVID et al., 1994)
◇   Fontainebleau sandstone (DAVID et al., 1994)
+   hot-pressed calcite (BERNABÉ et al., 1982)
– · – · –   fitting power law with exponent 3 for hot-pressed calcite in the high porosity range

Figure 7

Permeability versus porosity measured on various sandstones during mechanical compaction by DAVID et al. (1994) and on calcite during chemical compaction by BERNABÉ et al. (1982). The permeability-porosity curves of Figure 6 are shown for comparison.

experiments in spite of the differences discussed above. Additionally, it should be appreciated that compaction in the laboratory differs from diagenetic compaction to some degree. In laboratory experiments the pressure increase is relatively fast and destructive processes prevail. During compaction over long geological times, however, additional processes occur, such as plastic deformation and crack healing, which both result in a smaller permeability reduction with decreasing porosity.

Another type of permeability-porosity relationship results from chemical compaction, i.e., combined pressure solution and cementation (BOURBIE and ZINSZNER, 1985; ZHU et al., 1995). These time-dependent processes were studied on calcite by BERNABÉ et al. (1982) in high-temperature laboratory experiments. The corresponding permeability and porosity data are plotted in Figure 7. For porosities larger than 0.13 permeability varies with porosity cubed; for porosities smaller than 0.13 permeability varies with porosity taken to a power of 10. In respect to the roughness and fractal dimension of the pore walls, these processes reduce the initial porosity and result in pore walls with smooth surfaces. Therefore r_{eff} is close to r_{site}, and c_1 in equation (7) is about 1. As a result, permeability varies with porosity cubed as described by equations (6) or (23). At later stages of chemical compaction the grains come into close contact, forming pores with high constrictivity. This means that c_1 decreases and assumes a value typical for average sandstones. Although the main process of chemical compaction is pressure solution, the resulting structural changes appear similar to those attributed to mechanical compaction with grain dislocation and crushing. The slope of 10 in the $k - \phi$ diagram implies a fractal dimension similar to that for average sandstones subjected to mechanical compaction.

Based on measurements on different rock types as well as on high-pressure, high-temperature laboratory experiments, we conclude that during their diagenetic evolution sediments may follow one of two diagenetic paths. (1) In the general case pure mechanical compaction prevails during the first stage which is characterized by a large slope of the $k - \phi$ diagram. At a later stage this slope becomes smaller in the porosity range 0.1–0.2 for average sandstones which indicates that other processes become dominating. Microscopical investigations indicate chemical compaction for instance by the precipitation of quartz. This concept is confirmed by results obtained by BJØRKUM et al. (1998) who employed a complex numerical model to explain the variation of porosity with depth and time by mechanical compaction and quartz dissolution and precipitation. (2) The second, less frequent, diagenetic path is characterized by initial cementation and pressure solution, such as in Fontainebleau sandstone. In this case the $k - \phi$ diagram starts with a slope of three. Later the slope becomes larger, but in contrast to the first path, due to chemical compaction only.

Conclusions

In the Kozeny-Carman equation permeability varies with the square of the effective hydraulic pore radius r_{eff}. Based on a fractal model for porous media, r_{eff} can be expressed by the specific surface or by the grain radius and porosity taken to broken powers. This differs from the simple relationships derived for models with smooth capillaries or with smooth grains. A comparison of the fractal relationships with data from sedimentary rocks shows that they yield far better results than the simple relationships.

When the fractal relationships for the effective pore radius are inserted into the Kozeny-Carman equation, accordingly modified expressions for permeability are obtained. These can be used for calculating permeability from borehole measurements of porosity and formation factor.

The results are confirmed both by laboratory measurements on core samples and by high-pressure and high-temperature laboratory experiments. The fractal model can also be applied to interpret the diagenetic evolution of the pore space in respect to porosity and permeability caused by mechanical and chemical compaction during burial of sediments over geologic time.

Acknowledgements

The research reported in this paper was supported by the German Federal Ministry for Education, Science, Research, and Technology (BMBF) under grant 032 69 95. H. Kern (Institute of Geosciences, University of Kiel, Germany) is gratefully acknowledged for his constructive comments which helped us to improve the manuscript.

Appendix A

In this study a fractal pore space model was used to derive an expression (equation 18) for the path elongation Λ of a tortuous hydraulic path. This approach may shed light on the question whether the geometrical tortuosity is equal to Λ or to Λ^2. In the literature this topic is discussed with controversy.

In our definition of tortuosity T (equation 16) we follow CARMAN (1937) who demonstrated that $T \propto \Lambda^2$ for a model with smooth tortuous capillaries. If the wall of the capillaries has a fractal structure, then the elongation of the hydraulic path inside the capillary is smaller than the elongation Λ_{max} of a path which follows the pore wall. Λ_{max} can be calculated from equation (18) after substituting the fractal dimension D of the pore wall for D' the fractal dimension of the hydraulic capillary. If it can be assumed, as in equation (19), that $2 - D' = (2 - D)/2$, then equation (20) yields $\Lambda_{max} = \Lambda^2$. Thus, the tortuosity may be proportional to the linear path elongation along the pore wall for special geometries with a certain degree of wall roughness, i.e., in case that the elongation of a path along the wall is just the square of the elongation of a smoother path in the center of the capillary.

References

Archie, G. E. (1942), *Electrical Resistivity as an Aid in Core Analysis Interpretation*, Trans. AIME *146*, 54–61.

Bernabé, Y., Brace, W. F., and Evans, E. (1982), *Permeability, Porosity and Pore Geometry of Hot-pressed Calcite*, Mech. Mater. *1*, 173–183.

Bjørkum, P. A., Oelkers, E. H., Nadeau, P. H., Walderhaug, O., and Murphy, W. M. (1998), *Porosity Prediction in Quartzose Sandstones as a Function of Time, Temperature, Depth, Stylolite Frequency, and Hydrocarbon Saturation*, AAPG Bull. *82* (4), 637–648.

Brunauer, S., Emmett, V. H., and Teller, E. (1938), *Adsorption of Gases in Multimolecular Layers*, J. Am. Chem. Soc. *60*, 309–319.

Bourbie, T., and Zinszner, B. (1985), *Hydraulic and Acoustic Properties as a Function of Porosity in Fontainebleau Sandstone*, J. Geophys. Res. *90* (B13), 11,524–11,532.

Carman, P. C. (1937), *Fluid Flow through Granular Beds*, Trans. Inst. Chem. Eng. London *15*, 150.

Carman, P. C. (1948), *Some Physical Aspects of Water Flow in Porous Media*, Discuss. Faraday Soc. *3*, 78.

Carman, P. C., *Flow of Gases through Porous Media* (Butterworth Scientific Publications, London 1956).

David, C. (1993), *Geometry of Flow Paths for Fluid Transport in Rocks*, J. Geophys. Res. *98* (B7), 12,267–12,278.

David, C., Gueguen, Y., and Pampoukis, G. (1990), *Effective Medium Theory and Network Theory Applied to the Transport Properties of Rocks*, J. Geophys. Res. *95* (B5), 6993–7005.

David, C., Wong, T., Zhu, W., and Zhang, J. (1994), *Laboratory Measurements of Compaction-induced Permeability Change in Porous Rocks: Implications for the Generation and Maintenance of Pore Pressure Excess in the Crust*, Pure appl. geophys. *143* (1/2/3), 425–456.

Debschütz, W. (1995), *Hydraulische Untersuchungen an Sediment- und Kristallingesteinen unter variablen hydro- und lithostatischen Druckbedingungen: Trennung strömungs-charakterisierender Kenngrößen und Korrelation mit anderen petrophysikalischen Größen*, Ph.D. Dissertation Techn. Univ. Clausthal, Clausthal-Zellerfeld.

Iffland, J., and Voigt, H.-D. (1996), *Porositäts- und Permeabilitätsverhalten von Rotliegendsandsteinen unter Überlagerungsdruck*, DGMK-Berichte, Tagungsbericht 9602, Vorträge der Frühjahrstagung des DGMK-Fachbereiches Aufsuchung und Gewinnung, 25–26 April, Celle, Deutsche wissenschaftliche Gesellschaft für Erdöl, Erdgas und Kohle, e.V. (DGMK).

Kozeny, J. (1927), *Über die kapillare Leitung des Wassers im Boden (Aufstieg, Versickerung und Anwendung auf die Bewässerung)*, Sitz. Ber. Akad. Wiss. Wien. Math. Nat. (Abt. IIa) *136a*, 271–306.

Kulenkampff, J. (1994), *Die komplexe elektrische Leitfähigkeit poröser Gesteine in Frequenzbereich von 10 Hz bis 1 Mhz—Einflüsse von Porenstruktur und Porenfüllung*, Ph.D. Dissertation Techn. Univ. Clausthal, Clausthal-Zellerfeld.

Michaels, A. S., and Lin, C. S. (1954), *Permeability of Kaolinite*, Ind. Eng. Chem. *45*, 1239–1246.

Pape, H., and Schopper, J. R., *Relations between Physically Relevant Geometrical Properties of a Multifractal Porous System*. In *Characterization of Porous Solids* (ed. Unger, K. K. *et al.*) (Elsevier, Amsterdam 1988) pp. 473–482.

Pape, H., Clauser, C., and Iffland, J. (1999), *Permeability Prediction Based on Fractal Pore Space Geometry*, Geophysics *64* (5), 1447–1460.

PAPE, H., RIEPE, L., and SCHOPPER, J. R. (1982) *A Pigeon-hole Model for Relating Permeability to Specific Surface*, Log Analyst 23, 1, 5–13; Errata, Log Analyst, 23, 2, 50.

PAPE, H., RIEPE, L., and SCHOPPER, J. R. (1984), *The Role of Fractal Quantities, as Specific Surface and Tortuosities, for Physical Properties of Porous Media*, Particle Characterization 1, 66–73.

PAPE, H., RIEPE, L., and SCHOPPER, J. R. (1987a), *Theory of Self-similar Network Structures in Sedimentary and Igneous Rocks and their Investigation with Microscopical Methods*, J. Microscopy 148, 121–147.

PAPE, H., RIEPE, L., and SCHOPPER, J. R. (1987b), *Interlayer Conductivity of Rocks—A Fractal Model of Interface Irregularities for Calculating Interlayer Conductivity of Natural Porous Mineral Systems*, Colloids and Surfaces 27, 97–122.

SCHLÖMER, S., and KROOSS, B. M. (1997), *Experimental Characterization of the Hydrocarbon Sealing Efficiency of Cap Rocks*, Marine and Petrol. Geol. 14 (5), 565–580.

SCHOPPER, J. R. (1967), *Experimentelle Methoden und eine Apparatur zur Untersuchung der Beziehungen zwischen hydraulischen und elektrischen Eigenschaften loser und künstlich verfestigter poröser Medien*, Geophys. Prospecting 15 (4), 651–701.

VAN BRAKEL, J. (1975), *Pore Space Models for Transport Phenomena in Porous Media—Review and Evaluation with Special Emphasis on Capillary Liquid Transport*, Powder Technol. 11, 205–236.

ZHU, W., DAVID, C., and WONG, T. (1995), *Network Modeling of Permeability Evolution during Cementation and Hot Isostatic Pressing*, J. Geophys. Res. 100 (B8), 15,451–15,464.

(Received June 5, 1998, revised November 11, 1998, accepted December 12, 1998)

To access this journal online:
http://www.birkhauser.ch

Slow Two-phase Flow in Single Fractures: Fragmentation, Migration, and Fractal Patterns Simulated Using Invasion Percolation Models

GERI WAGNER,[1,2] HÅKON AMUNDSEN,[1] UNNI OXAAL,[3] PAUL MEAKIN,[1] JENS FEDER[1] and TORSTEIN JØSSANG[1]

Abstract—The slow displacement of a wetting fluid by an invading non-wetting fluid in single fractures was studied using experiments and simulations. In the experiments, the fracture aperture was modeled by the gap between a rough plate and a smooth transparent plate. The displacement was simulated using invasion percolation models and two types of self-affine fracture aperture models; model A with an infinite in-plane correlation length, and model B with a finite in-plane correlation length. Simulations were also performed on self-affine models that precisely represented the aperture fields of the experiments. At length scales below the in-plane correlation length, the simulated displacement patterns show scaling properties that may be tuned by changing the characteristics of the underlying geometry. In the experiment-matched simulations, we observed closely corresponding displacement patterns.

Key words: Fracture, two-phase flow, invasion percolation.

1. Introduction

The slow displacement of one fluid by another, immiscible fluid in porous and fractured rock is difficult to describe due to the many different length scales associated with the heterogeneities in rock. A necessary step towards a better understanding of flow in systems such as fractured rocks in oil reservoirs and aquifers is to study patterns in single fractures. In this paper we present the results of computational and experimental studies of slow immiscible fluid-fluid displacement processes in fracture models. Displacements of this kind are dominated by capillary forces and are important in processes such as migration of oil from source to reservoir rocks (ENGLAND *et al.*, 1987; DEMBICKI, Jr. and ANDERSON, 1989; BETHKE *et al.*, 1991).

---

[1] Department of Physics, University of Oslo, Box 1048, Blindern, 0316 Oslo 3, Norway.
[2] Present address: Raymond and Beverly Sackler Faculty of Exact Sciences, School of Physics and Astronomy, Tel Aviv University, Ramat Aviv 69978, Tel Aviv, Israel.
[3] Department of Agricultural Engineering, Agricultural University of Norway, PB 5065, N-1432 Ås, Norway.

The geometry of rock fractures has been studied in great detail. The spatial variations of fracture apertures are strongly correlated: the points of contact or small apertures are likely to be surrounded by other points of contact or small apertures, while points of large apertures are likely to be surrounded by other points of large aperture. MANDELBROT et al. (1984) showed that the surfaces of fractured steel specimens could be described in terms of fractal geometry. BROWN (1987, 1988a) and ODLING (1994) reported that self-affine fractal models provided good descriptions of natural fracture surfaces. POWER et al. (1987, 1991) and SCHMITTBUHL et al. (1993a) also found that fractured rock surfaces exhibited self-affine scaling properties. SCHMITTBUHL et al. (1995) studied fresh brittle fracture surfaces of granite and gneiss and found self-affine fractal structures.

Fractional (or fractal) Brownian motion (fBm) is a mathematical construction that is used to generate self-affine fractals (FEDER, 1988). In one dimension, the Brownian motion $B(t)$ describes the position at time t of a particle that takes steps in either direction with equal probability. fBm is a generalization of Brownian motion in which the motion is either persistent or anti-persistent. In persistent motion, steps in one of the two directions are more likely to be followed by another step in the same direction, whereas in anti-persistent motion, they are more likely to be followed by a step in the opposite direction. The cited studies of fractured rock surfaces motivated the use of fBm surfaces in this work to construct fracture models. In our opinion, the models may serve as prototypes of natural fractures. Fluid-fluid displacement processes in these fracture models were studied experimentally and by computer simulations. In both experiments and simulations, a scenario was considered in which a wetting fluid (usually water in geological applications) initially filling a fracture void space is slowly displaced by a non-wetting fluid (such as oil, air or non-aqueous polluting liquids).

Our simulations were based on the invasion percolation (IP) model (LENORMAND and BORIES, 1980; CHANDLER et al., 1982; WILKINSON and WILLEMSEN, 1983). Percolation models are frequently used in the context of pore-scale flow through porous media, and the fluid-fluid displacement patterns obtained from these models are often in very good agreement with experimental observation (see, for example, LENORMAND and ZARCONE, 1985). Here, the IP model was used to simulate the quasistatic displacement of a wetting fluid by an invading non-wetting fluid in a fracture. We studied variable-aperture fracture models in which the rough surfaces of a fracture were approximated using a square lattice in which each site represented a region of the fracture with a more or less constant aperture, and the aperture varied from site to site (PATIR and CHENG, 1978; BROWN, 1988b, 1989; MORENO et al., 1988). In section 2, IP model simulations are presented and analyzed quantitatively, using two simple fracture aperture models based on fBm. Although invasion percolation models are highly successful in modeling flow through pores, it is not clear to what degree these models capture real-world displacement processes in single fractures. In section 3, we verify the assumptions used in the modeling by directly comparing simulations and experiments of slow immiscible displacements. A brief summary is given in section 4.

2. Simulations

Given the fractal character of fresh fracture surfaces, the most simple conceptual model for a fracture aperture is the space between two self-affine fBm surfaces that are oriented parallel to each other. In some regions, the surfaces may overlap. These regions represent occluded parts of the fracture aperture in which the two surfaces come into contact and no flow is possible. In the remaining regions, the surfaces are separated from each other and form a void space with both wide and narrow voids.

Fractal Brownian surfaces appear rough (fractal) on short length scales or distances, and flat (Euclidean) on long length scales. A self-affine surface $z(x, y)$ is characterized by the scaling property

$$\langle |z(x_1, y_1) - z(x_2, y_2)| \rangle \sim d_{12}^H, \tag{1}$$

where (x_1, y_1) and (x_2, y_2) denote two points in the surface plane and $d_{12} = [(x_2 - x_1)^2 + (y_2 - y_1)^2]^{1/2}$ is the distance between these points. The symbol $\sim$ should be interpreted as meaning "scales as" and $\langle \cdots \rangle$ indicates averaging over many pairs of points separated by the same distance $d = d_{12}$. The Hurst exponent or roughness exponent H varies from 0 to 1 and characterizes the surface height-height correlations. In a fBm surface, height correlations exist on all length scales; low values of H ($H < 1/2$) indicate anti-persistance and high values of H ($H > 1/2$) indicate persistence of the fluctuations in the topography.

A. Model A: fBm Surface and Planar Surface

If the two fBm surfaces $z_1(x, y)$ and $z_2(x, y)$ are characterized by equal Hurst exponents H and equal amplitudes, then the aperture field $b(x, y)$ given by $b(x, y) = z_1(x, y) - z_2(x, y)$ is also a self-affine function characterized by the same exponent H (WANG et al., 1988; PLOURABOUE et al., 1995a). Consequently statistically identical void space geometries with fractal properties are obtained by using either the void space between two self-affine surfaces, or the void spaces between a self-affine surface and a planar surface. Figure 1 illustrates this fracture model.

In a fracture, the capillary pressure ΔP_c opposing the displacement of a wetting fluid by an invading non-wetting fluid is inversely proportional to the local aperture $b(x, y)$, $\Delta P_c = 2\gamma \cos \theta / b$, where θ is the wetting angle and γ is the surface tension that characterizes the interface between the displacing fluid and the displaced fluid. This relationship can be derived from the more general Young-Laplace equation

(KUEPER and MCWHORTER, 1991), $\Delta P_c = \gamma(R_1^{-1} + R_2^{-1})$, where R_1 and R_2 denote the principal radii of curvature of the interface. If it is assumed that the interface is nearly flat in the direction parallel to the fracture plane ($R_1 \gg R_2$), the Young-Laplace equation can be simplified by replacing R_2^{-1} by $2\cos\theta/b$ and R_1^{-1} by 0. Physically, the equation expresses the fact that the non-wetting fluid easily enters regions with wide apertures, but it cannot displace the wetting fluid from regions with narrow apertures.

In the IP model, random numbers ("invasion thresholds") are assigned to each site on a lattice representing a random medium. Initially, all but one of the sites are occupied by the wetting fluid ("defender fluid"), and one site, the injection site, is occupied by the non-wetting fluid ("invader fluid"). The simulation proceeds stepwise. In each step, the defender fluid site adjacent to the invaded region that has the *lowest* invasion threshold is filled with invader fluid. The simulation is terminated when the invading fluid reaches the lattice boundary. As a result, the displacement process is described at the pore level, and a detailed representation of all stages of the fluid-fluid displacement process is obtained. Viscous forces are ignored in this model. A convenient procedure for simulating slow displacement processes in a fracture (here called fracture model A) is:

1. Generate a self-affine fBm surface $z(x, y)$.

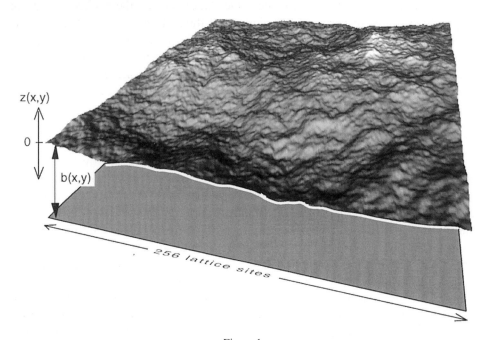

Figure 1
Fracture model A is formed by a self-affine fBm surface $z(x, y)$ and a planar surface. The fBm surface shown here is characterized by $H = 0.8$.

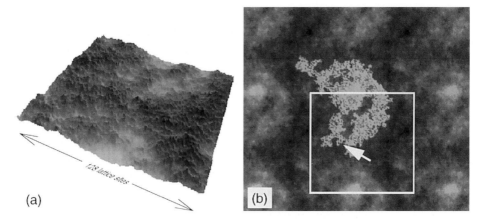

Figure 2
IP cluster (with trapping) grown on a self-affine topography (fracture model A) using $H = 0.5$. Part (a) shows a representation of the fBm surface used in the simulation, and part (b) shows the cluster (white) grown on the corresponding substrate. The gray scale indicates the elevation $z(x, y)$ of the surfaces and the invasion thresholds p_i of the ith site of the substrates. Dark shades represent regions with high elevations and low thresholds and light-gray shades represent regions with low elevations and high thresholds. In the simulations, the substrates were extended periodically. The outline of the substrate and the injection site is indicated (arrow).

2. To each site on a lattice with the same dimensions as the surface, assign an invasion threshold p_i given by the inverse of the surface elevation $z(x_i, y_i)$ at the ith site, $p_i = z^{-1}(x_i, y_i)$, where (x_i, y_i) is the position of the ith site. [Because the IP model depends only on the relative magnitudes of the thresholds, but not on their absolute magnitudes, a statistically equivalent pattern is obtained by setting $p_i = z(x_i, y_i)$. This is no longer true for invasion percolation models in the presence of a gradient, discussed later in this article.] If $z(x_i, y_i)$ is negative, p_i is set to infinity to represent an occluded region.
3. Carry out the simulation using the invasion percolation algorithm.

Figure 2 shows a displacement pattern obtained in this manner, using $H = 0.5$. The fBm surface was generated using a random midpoint displacement algorithm with random successive addition (VOSS, 1985; SAUPE, 1988). All fBm surfaces used in our work were generated using a Gaussian distribution of displacements and were constructed on square lattices of size $L \times L$. In order to maximize the size of the IP clusters, periodic boundary conditions were used in the simulations, the fBm surfaces were extended periodically, and each simulation was terminated when the growing IP cluster intersected itself. It was apparent that the cluster preferred to grow in regions of the surfaces with high elevations, or large apertures, where the invasion thresholds were low. A trapping rule (WILKINSON and WILLEMSEN, 1983) was used. This rule conserved the volume of "islands" of defender fluid that were not connected to the lattice boundary by a path consisting of steps between nearest-neighbor defender fluid sites, and represented the essentially incompressible nature of the defender fluid. These "islands" could not be invaded by the invader fluid, even if their invasion thresholds were favorable.

The IP clusters grown in fracture model A could be described as a collection of "blobs" of invading fluid, connected to each other by fine "threads." This morphology became more distinct as H was increased. IP clusters of this kind were studied recently by PATERSON et al. (1996a, 1996b) and by DU et al. (1995, 1996). In our own investigations (WAGNER et al., 1997), we found that the IP clusters had fractal properties that could be tuned by varying the Hurst exponent H characterizing the void space geometry. Table 1 lists the fractal dimensionalities measured for different values of H, measured on fBm surfaces with a size of 512×512 lattice units. The fractal dimensionality D of IP clusters increases (the clusters become more compact) as the degree of persistence characterizing the void space increases. Measurements of D are potentially useful for comparison with experiments, but there is also theoretical interest in IP on self-affine topographies (ISICHENKO, 1992; SCHMITTBUHL et al., 1993b; SAHIMI, 1995). For comparison, in IP with trapping on uncorrelated random substrates (corresponding to $H = 0$) the fractal dimensionality of IP clusters is approximately 1.82 (FURUBERG et al., 1988).

B. Model B: Gap between two Displaced Identical fBm Surfaces

Studies of natural fracture surfaces indicate that the geometries of the two bounding surfaces of a natural fracture aperture tend to be correlated (BROWN et al., 1986). In fracture model A, these surface cross-correlations cannot be represented. WANG et al. (1988) proposed a single fracture aperture model that reproduces the correlations between the bounding surfaces. In this model (fracture model B), the two bounding surfaces of the fracture aperture are represented by a self-affine fBm surface $z(x, y)$ and a replica $z_c(x, y)$ of this surface, which is oriented parallel to the original and displaced relative to the original by a distance b perpendicular to the fracture plane and by a translation vector $\mathbf{r_d}$ parallel to the fracture plane. "Periodic wrapping" is used in the lateral displacement so that the top surface covers the entire bottom surface. The fracture plane is given by the orientation of the surfaces. The aperture field is given by the elevation difference,

Table 1

List of fractal cluster dimensionalities D measured in simulations using fracture model A and various Hurst exponents H characterizing the fracture void space. The fBm surfaces used had sizes of 512×512 lattice units.

H	D
0.31	1.89 ± 0.01
0.47	1.91 ± 0.01
0.85	1.94 ± 0.01

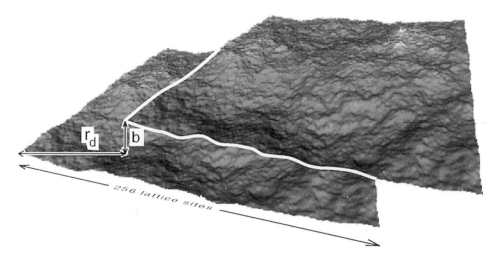

Figure 3
Fracture model B consisting of a self-affine fBm surface and its replica, displaced by a distance b in the direction perpendicular to the fracture plane and by a distance $r_d = |\mathbf{r}|_d$ parallel to the fracture plane.

$b(x, y) = z(x, y) - z_c(x, y)$. In the absence of surface overlapping, the distance b corresponds to the mean aperture of the fracture. If the two surfaces overlap, the aperture is set to zero. Even using a value of $b = 0.0$, a void space with local positive elevation differences $b(x, y)$ will be generated if a nonzero offset distance r_d is used. Figure 3 illustrates the procedure.

The properties of the void spaces between fBm surfaces were studied by WANG et al. (1988) and ROUX et al. (1993) and more recently by PLOURABOUE et al. (1995a, 1995b). For a given fracture obtained by displacing a self-affine surface and its replica, the offset distance r_d in the direction of the fracture plane plays the role of a correlation length. The aperture field has the properties of a self-affine fractal on length scales less than r_d. The scaling properties of the aperture field and of the bounding surfaces are characterized by the same exponent H. On length scales greater than r_d, the spatial correlations are lost so that the apertures at two points that are separated by a distance larger than r_d are independent of each other.

In model B, the fracture was characterized by three parameters, H, r_d, and b. In our work (WAGNER et al., 1999) we focused on a Hurst exponent of $H = 0.8$. This value of H is consistent with an experimental study of crack surfaces in rock (SCHMITTBUHL et al., 1995). Depending on the model parameters, the asperities of a surface and its replica could overlap in some regions, leading to negative differences. The aperture field $b(x, y)$ was taken to be zero in those regions and represented occluded regions. The invasion threshold p_i of a site was given by the inverse of the aperture, $p_i = b^{-1}(x_i, y_i)$.

We found that the geometric cross-over of the aperture field was reflected by the structure of the IP cluster: On length scales greater than r_d, the clusters had scaling properties similar to those of ordinary IP clusters grown on uncorrelated substrates. On shorter length scales, the cluster structures shared the properties of the IP clusters observed in model A. Figure 4 shows an IP cluster obtained in a typical simulation. Note that length-scale dependence of cluster structure cannot be appreciated from the figure, it must be extracted by analyzing a series of simulations.

Characteristic length scales may be imposed on the displacement patterns not only by void space correlations but also by external forces, such as a gravity. Figure 5(a) shows an IP cluster obtained in a simulation in which the non-wetting and less-dense fluid entered at the lower fracture edge. Periodic boundary conditions were used in the lateral direction, and the simulation was terminated when the non-wetting fluid had reached the top edge. The parameters used were $H = 0.8$, $r_d = 128$, and $b = 0.5$. Figure 5(b) shows a cluster obtained by using the same void space and imposing a gradient $g = 0.01$ on all the invasion thresholds $p_i(x, y)$. The contribution $\Delta p_i = -gy_i$ was added to all invasion thresholds $[p_i(x_i, y_i) \to p_i(x_i, y_i) - gy_i]$. This contribution represented the effect of gravity acting due to a density different between the two fluids so that the invading (less dense) non-wetting fluid tended to invade regions towards the top of the fluid-fluid interface. The resulting IP cluster is more slender than its counterpart at zero gravity. IP in homogeneous random porous media in the presence of gravity has been studied in detail (WILKINSON, 1984; MEAKIN et al., 1992). From these studies it is well known that gravity does not change the IP cluster structure but limits the scaling behavior

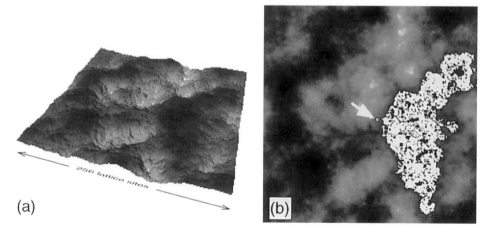

Figure 4
IP cluster grown on a correlated substrate (fracture model B), using the parameters $H = 0.8$, $b = 0.0$, and $r_d = 32$ (measured in lattice units). Part (a) shows a representation of the aperture field used in the simulation, and part (b) shows the cluster (white) grown on the corresponding substrate. The gray scale indicates the aperture, with dark shades representing regions with large apertures and light-gray shades representing regions with narrow or zero apertures. The injection site is indicated by an arrow.

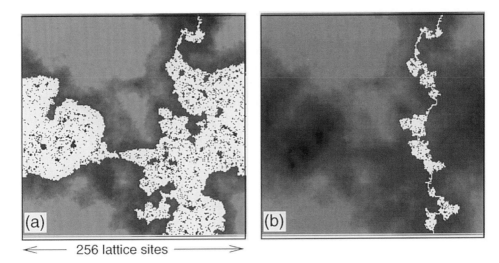

Figure 5
IP cluster grown on a correlated substrate (fracture model B), using the parameters $H = 0.8$, $b = 0.5$, and $r_d = 128$ (measured in lattice units) with a gradient g imposed on the invasion thresholds. Part (a) shows the cluster (white) obtained in a horizontal model ($g = 0.0$), and part (b) shows the cluster obtained in a tilted model ($g = 0.01$). The gray scale indicates the apertures, with light-gray shades representing regions with narrow or zero apertures. The non-wetting and less-dense fluid entered at the lower edge, and periodic boundary conditions were used in the lateral direction.

to a characteristic length, or correlation length, ξ that depends on the magnitude of g. However, we have not investigated quantitatively the effect of gravity on fluid-fluid displacements in fracture apertures.

3. Experiments

Both fracture models considered in the previous section are highly idealized. The IP model is also an idealization, based on the assumption that the dominating forces are given by the local apertures only. It is not clear if these simple models are justified. For example, the IP representation of the displacement process ignores the effects of matrix erosion and deposition, thermal fluctuations, heterogeneous and changing wetting properties, and the three-dimensional nature of the fracture. To investigate applicability of the model, we focus on the representation of displacement processes by IP, and compare the simulations with experiments.

Our experiments were based on fracture model A. A single fracture was represented by the gap between a rough plate and a planar transparent plate, both oriented horizontally and clamped together. Two different rough plates were used. In model A_1, a PMMA plate with a pattern of 40×40 sites, each with an area of 5×5 mm milled into it, was used. In model A_2, a textured glass plate was used. The aperture fields of both models ranged from 0 to about 0.3 mm.

In each experiment, the fracture models were carefully cleaned and filled with water containing 0.1 weight percent Nigrosine (a black dye) and placed on a light box. Air was injected slowly through an inlet in the center of the planar top plate, at a rate of 0.1 ml/min. Water was collected at the boundary of the models. The experiment was terminated when air reached the boundary. In both models, top plates of the same material as the bottom plates were used to create uniform wetting conditions. Details of the experimental procedure and of the simulation models used for comparison can be found in a separate publication (AMUNDSEN et al., 1999).

A. Model A_1: fBm Surface and Planar Surface

The surface pattern of the PMMA bottom plate used in model A_1 was machined to correspond to a computer-generated fBm surface characterized by a Hurst exponent of $H = 0.8$. The fBm surface was discretized on 1600 sites, using 30 different depths ranging between 0.0 and 0.3 mm. A planar PMMA top plate was placed in contact with this machined surface to obtain a realization of fracture model A discussed in section 2A. The fluid-fluid displacement experiment was carried out in the gap between the two plates.

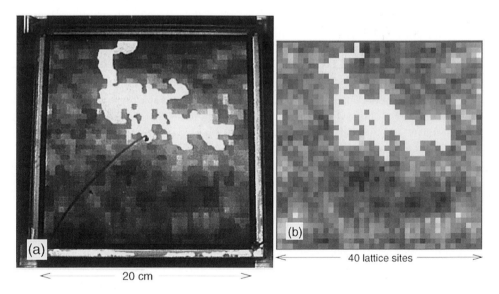

Figure 6

Part (a) shows the final stage in a displacement experiment using model A_1. Air (white) was injected in the center of the fracture model. Part (b) shows the IP cluster (white) obtained in a simulation. The gray scale indicates the aperture, with dark shades representing regions with large apertures and light shades representing regions with small or zero apertures.

Figure 6(a) shows the final stage of a displacement experiment using this model. As in the previous section, regions with wide apertures appear dark (due to strong light absorption), and the invading air is shown in white. The air formed a connected structure that covered the sites with the greatest apertures. Small trapped islands of water may be observed. At some stages during the cluster growth, several water-filled sites could be invaded that had apertures of the same nominal size. At these degenerate configurations, the cluster growth was determined by impurities in the model or by inaccuracies in the model geometry. As a result of those degeneracies, the cluster of air did not grow in a completely reproducible manner in different experiments.

Figure 6(b) displays an IP cluster obtained in a simulation of this displacement experiment. In the simulation, the void space data employed to construct the experimental fracture model was used. Air injection was simulated by non-wetting fluid invading a site in the center of the lattice. At degenerate states at which several sites with equally low invasion threshold could be invaded, all of those sites were invaded. No trapping rule was used.

From Figure 6, it is apparent that the simulated displacement pattern captures important features of the experimental pattern. Experimental simulated displacement patterns were compared site by site, at all stages of the displacement. The general agreement was good, with more than 70% overlap during most of the invasion process. The scaling properties of the experimental clusters could not be studied due to the very small system size.

B. Model A_2: Textured Glass Plate and Planar Surface

The void space formed by a textured glass plate and a planar glass plate was anisotropic and characterized by deep "valleys" and shallow "ridges." On short length scales, the aperture field has properties similar to a self-affine fBm surface with a Hurst exponent of $H = 0.82 \pm 0.08$. The experimental model had a size of 30×30 cm, and the apertures varied continuously, unlike the stepwise changes in model A_1. Apertures were measured employing a light-absorption technique (NICHOLL and GLASS, 1994) and found to range from 0 to 0.35 mm, with an approximately log-normal distribution.

Figure 7(a) shows a displacement pattern observed at the final stage. During the invasion process, the air fragmented into numerous clusters, displaying interesting dynamics. Fragmentation occurred rapidly on the time scale of the experiment. The single fragments migrated to regions with wide apertures and remained there until further fragments coalesced, forming a larger fragment. Eventually the larger fragments could re-distribute their fluid contents to an even wider region, undergoing further fragmentation on their way. At no stage of the experiment did the air form a connected path from the inlet to the boundary of the fracture model.

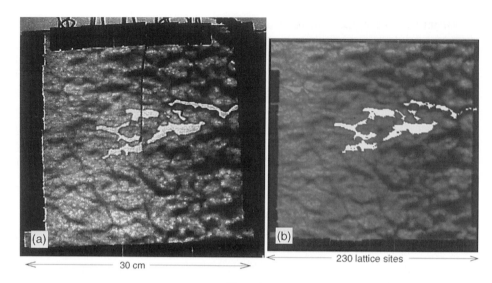

Figure 7
Part (a) shows the final stage in a displacement experiment using model A_2. Air (white) was injected in the center of the fracture model. Part (b) shows the structure (white) obtained in a simulation using a modified IP model. The gray scale indicates the aperture, with dark shades representing regions with large apertures and light shades representing regions with small or zero apertures.

No trapping occurred in model A_2. This fact and the occurrence of fragmentation and fluid-redistribution was attributed to the pronounced anisotropy of the void space geometry and to the continuous aperture variations (AMUNDSEN et al., 1999). Repeated experiments produced very similar displacement patterns.

The displacement was simulated using a modified IP model that included fragmentation and fluid-redistribution processes. Figure 7(b) shows the final stage of a simulation performed on the same aperture field as used in the experiment. The aperture field of the experimental fracture was measured using light-adsorption measurements and was discretized using a square lattice of 230×230 sites. During the simulation, an IP cluster was formed around the injection site in the center of the lattice, using the standard IP algorithm with trapping. At each stage, redistribution of invader fluid occurred if a cluster was adjacent to a defender fluid site with a larger aperture than one of the cluster sites at the perimeter of the cluster. In this case, the fluid contents of the two sites were swapped. This could lead to fragmentation of the IP cluster and simulate migration of a fragment in the void space. In this algorithm, the projected fluid area (the number of invaded sites) was conserved, rather than fluid volume.

Again, experimental and simulated displacement patterns were carefully compared. The simulation produced structures that were surprisingly similar to the experimental ones (with better than 80% overlap during most of the displacement) and also captured important aspects of the fragmentation dynamics. The good agreement may be related to the fact that the aperture field was continuous and, unlike model A_1, no aperture degeneracies were present.

4. Summary

The intrinsic correlations of the rock matrix make two-phase flow in fractures an interesting problem to study, because the resulting fluid structures both reflect the dynamics of the displacement process and the morphology and topology of the fracture network. In this work we studied flow through single fractures. Simulations of displacement processes in fractures of type A, consisting of the void space between one fBm surface and one plane surface, exhibit a fractal scaling behavior that is systematically dependent on the surface roughness, see Table 1. Simulations on model B type fractures, consisting of a fBm surface and its replica displaced by an in-plane vector r_d, show a cross-over phenomenon. Below the correlation length set by r_d, the displacement structures scale like those on model A; above this length scale, the structures have the properties of ordinary IP clusters grown on uncorrelated substrates.

From the experiments and simulations discussed in section 3, we conclude that IP and modified IP models can be useful to study fluid displacement patterns in single fractures. Fluid occupancy patterns can then be used to compute steady-state transport properties (relative permeabilities in two-phase flow) through a fracture, as in the work of MENDOZA and SUDICKY (1991).

The challenge of understanding flow and displacement in entire fracture networks in both porous and impermeable media remains. In these systems, the interplay between properties of single components, as studied in this work, and global network properties must be investigated.

Acknowledgments

We gratefully acknowlege support by VISTA, a research cooperation between the Norwegian Academy of Science and Letters and Den Norske Stats Oljeselskap A.S. (STATOIL) and by NFR, the Norwegian Research Council. The work presented has received support from NFR through a grant of computing time.

REFERENCES

AMUNDSEN, H., WAGNER, G., MEAKIN, P., FEDER, J., and JØSSANG, T. (1999), *Slow Two-phase Flow in Artificial Fractures: Experiments and Simulations*, Water Resour. Res. *35* (9), 2619–2629.

BETHKE, C. M., REED, J. D., and OLTZ, D. F. (1991), *Long-range Petroleum Migration in the Illinois Basin*, AAPG Bulletin 75 (5), 924–945.
BROWN, S. R. (1987), *A Note on the Description of Surface Roughness Using Fractal Dimension*, Geophys. Res. Lett. 14 (11), 1095–1098.
BROWN, S. R. (1988a), *Correction to "A note on the description of surface roughness using fractal dimension,"* Geophys. Res. Lett. 15 (3), 286.
BROWN, S. R. (1988b), *Fluid Flow through Rock Joints: The Effect of Surface Roughness*, J. Geophys. Res. 92 (B2), 1337–1347.
BROWN, S. R. (1989), *Transport of Fluid and Electric Current through a Single Fracture*, J. Geophys. Res. 94 (B7), 9429–9438.
BROWN, S. R., KRANZ, R. L., and BONNER, B. P. (1986), *Correlation between the Surfaces of Natural Rock Joints*, Geophys. Res. Lett. 13 (13), 1430–1433.
CHANDLER, R., KOPLIK, J., LERMAN, K., and WILLEMSEN, J. F. (1982), *Capillary Displacement and Percolation in Porous Media*, J. Fluid Mech. 119, 249–267.
DEMBICKI Jr., H., and ANDERSON, M. J. (1989), *Secondary Migration of Oil: Experiments Supporting Efficient Movement of Separate, Buoyant Oil Phase along Limited Conduits*, AAPG Bulletin 73(8), 1018–1021.
DU, C., SATIK, C., and YORTSOS, Y. C. (1996), *Percolation in a Fractional Brownian Motion Lattice*, AIChE Journal 42 (8), 2392–2395.
DU, C., XU, B., YORTSOS, Y. C., CHAOUCHE, M., RAKOTOMALALA, N., and SALIN, D. (1995), *Correlation of Occupation Profiles in Invasion Percolation*, Phys. Rev. Lett. 74 (5), 694–697.
ENGLAND, W. A., MACKENZIE, A. S., MANN, D. M., and QUIGLEY, T. M. (1987), *The Movement and Entrapment of Petroleum Fluids in the Subsurface*, J. Geological Society of London 144, 327–347.
FEDER, J., *Fractals* (Plenum Press, New York 1988).
FURUBERG, L., FEDER, J., AHARONY, A., and JØSSANG, T. (1988), *Dynamics of Invasion Percolation*, Phys. Rev. Lett. 61 (18), 2117–2120.
ISICHENKO, M. B. (1992), *Percolation, Statistical Topography, and Transport in Random Media*, Rev. Modern Physics 64 (4), 961–1043.
KUEPER, B. H., and McWHORTER, D. B. (1991), *The Behavior of Dense, Nonaqueous Phase Liquids in Fractured Rock and Clay*, Ground Water 29 (5), 716–728.
LENORMAND, R., and BORIES, S. (1980), *Description d'un mécanisme de connexion de liason destiné a l'étude du drainage avec piégeage en milieu poreux*, C.R. Acad. Sc. Paris 291, 279–283.
LENORMAND, R., and ZARCONE, C. (1985), *Invasion Percolation in an Etched Network: Measurement of a Fractal Dimension*, Phys. Rev. Lett. 54 (20), 2226–2229.
MANDELBROT, B. B., PASSOJA, D., and PAULLAY, A. J. (1984), *Fractal Character of Fracture Surfaces of Metals*, Nature 308, 721–722.
MEAKIN, P., FEDER, J., FRETTE, V., and JØSSANG, T. (1992), *Invasion Percolation in a Destabilizing Gradient*, Phys. Rev. A 46, 3357–3368.
MENDOZA, C. A., and SUDICKY, E. A., *Hierarchical scaling of constitutive relationships controlling multi-phase flow in fractured geologic media*. In *Reservoir Characterization: 3rd International Technical Conference: Papers* (ed. Linville, B.) (Pennwell, Tulsa OK 1991) pp. 505–514.
MORENO, L., TSANG, Y. W., TSANG, C.-F., HALE, F. V., and NERETNIEKS, I. (1988), *Flow and Tracer Transport in a Single Fracture: A Stochastic Model and its Relation to Some Field Observations*, Water. Resour. Res. 24 (12), 2033–2048.
NICHOLL, M. J., and GLASS, R. J. (1994), *Wetting Phase Permeability in a Partially Saturated Horizontal Fracture*, High-Level Radioactive Waste Management, American Nuclear Society, La Grange Park, IL, pp. 2007–2019.
ODLING, N. E. (1994), *Natural Fracture Profiles, Fractal Dimension and Joint Roughness Coefficients*, Rock Mech. Rock Eng. 27 (3), 135–153.
PATERSON, L., and PAINTER, S. (1996a), *Simulating Residual Saturation and Relative Permeability in Heterogeneous Formations*. Paper SPE 36523 presented at the Annual Technical Conference and Exhibition of the Society of Petroleum Engineers, held in Denver, CO., Oct. 6–9, 1996.
PATERSON, L., PAINTER, S., KNACKSTEDT, M. A., and PINCZEWSKI, W. V. (1996b), *Patterns of Fluid Flow in Naturally Heterogeneous Rocks*, Physica A 235, 619–628.

PATIR, N., and CHENG, H. S. (1978), *An Average Flow Model for Determining Effects of Three-dimensional Roughness on Partial Hydrodynamic Lubrication*, J. Lubr. Technol. *100*, 12–17.

PLOURABOUE, F., KUROWSKI, P., HULIN, J.-P., and ROUX, S. (1995a), *Aperture of Rough Cracks*, Phys. Rev. E *51* (3), 1675–1685.

PLOURABOUE, F., ROUX, S., SCHMITTBUHL, J., and VILOTTE, J.-P. (1995b), *Geometry of Contact between Self-affine Surfaces*, Fractals *3* (1), 113–122.

POWER, W. L., and TULLIS, T. E. (1991), *Euclidean and Fractal Models for the Description of Rock Surface Roughness*, J. Geophys. Res. *96* (B1), 415–424.

POWER, W. L., TULLIS, T. E., BROWN, S. R., BOITNOTT, G. N., and SCHOLZ, C. H. (1987), *Roughness of Natural Fault Surfaces*, Geophys. Res. Lett. *14* (1), 29–32.

ROUX, S., SCHMITTBUHL, J., VILOTTE, J.-PL., and HANSEN, A. (1993), *Some Physical Properties of Self-affine Rough Surfaces*, Europhys. Lett. *23* (4), 277–282.

SAHIMI, M. (1995), *Effect of Long-range Correlations on Transport Phenomena in Disordered Media*, AIChE Journal *41* (2), 229–240.

SAUPE, D., *Algorithms for random fractals*. In *The Science of Fractal Images* (H.-O. Peitgen and D. Saupe, eds.) (Springer-Verlag, Berlin 1988) pp. 71–136.

SCHMITTBUHL, J., GENTIER, S., and ROUX, S. (1993a), *Field Measurements of the Roughness of Fault Surfaces*, Geophys. Res. Lett. *20* (8), 639–641.

SCHMITTBUHL, J., VILOTTE, J.-P., and ROUX, S. (1993b), *Percolation through Self-affine Surfaces*, J. Phys. A: Math. Gen. *26*, 6115–6133.

SCHMITTBUHL, J., SCHMITT, F., and SCHOLZ, C. H. (1995), *Scaling Invariance of Crack Surfaces*, J. Geophys. Res. *100*, 5953–5973.

VOSS, R. F. (1985), *Random fractal forgeries*. In *Fundamental Algorithms for Computer Graphics* (R. A. Earnshaw, ed.) (Springer-Verlag, Berlin 1985) pp. 805–835.

WAGNER, G., MEAKIN, P., FEDER, J., and JØSSANG, T. (1997), *Invasion Percolation on Self-affine Topographies*, Phys. Rev. E *55* (2), 1698–1703.

WAGNER, G., MEAKIN, P., FEDER, J., and JØSSANG, T. (1999), *Invasion Percolation in Fractal Fractures*, Physica A *264*, 321–337.

WANG, J. S. Y., NARASIMHAN, T. N., and SCHOLZ, C. H. (1988), *Aperture Correlation of a Fractal Fracture*, Water Resour. Res. *93* (B3), 2216–2224.

WILKINSON, D. (1984), *Percolation Model of Immiscible Displacement in the Presence of Buoyancy Forces*, Phys. Rev. A *30* (1), 520–531.

WILKINSON, D., and WILLEMSEN, J. F. (1983), *Invasion Percolation: A New Form of Percolation Theory*, J. Phys. A: Math. Gen. *16*, 3365–3376.

(Received April 6, 1998, revised October 28, 1998, accepted December 12, 1998)

To access this journal online:
http://www.birkhauser.ch

Dynamic Model of the Infiltration Metasomatic Zonation

V. L. RUSINOV[1] and V. V. ZHUKOV[1]

Abstract—A new dynamic model of infiltration metasomatic zonation is discussed. The model describes the formation and evolution of metasomatic zoning patterns in terms of travelling fronts and waves. Travelling fronts cause the formation of well known vectoral zonation. All replacement reactions in this zonation are localized immediately around fronts (zone boundaries), and inside individual zones the chemical equilibrium is attained. Such a condition corresponds to the "local equilibrium model" and no regular oscillations occur. Travelling waves can stimulate stable spatial chemical oscillations and appearance of the dissipative structure in the solution flow, if redox reactions such as $Fe^{2+} \leftrightarrow Fe^{3+} + e^-$ in solution occur. Corresponding to the evolution of the dissipative structure, the rhythmical banding in the metasomatic rocks is formed. Thus the proposed model describes the single approach as vectoral as periodic (rhythmically banded) zoning patterns.

Key words: Metasomatic zonation, travelling waves, dynamic model, dissipative structure.

Introduction

Infiltration metasomatic zonation is a kind of spatial ordering of mineral assemblages due to the chemical interaction between rocks and solution flow filtrating through it. The main parameters driving the mineral formation in such systems are: initial component concentrations in the solution, initial mineral composition of the rock, rate of filtration, chemical reaction rates, and temperature. We limit the further discussion with isothermal systems showing no spatial T-gradient, and preserving constant volume. Therefore remaining independent parameters in the systems under discussion are initial component concentrations, filtration rate, and rates of chemical reactions. If the solution does not react with the host rock then the dynamics of mass exchange (dm_i) in the system depends only on spatial coordinate (x), concentration of component (C_i), and filtration coefficient value (φ_i)

$$dm_i = \varphi_i C_i \, dV,$$

---

[1] Institute of Geology of Ore Deposits, Petrography, Mineralogy and Geochemistry, Russian Academy of Science, Staromonetnyi, 35, Moscow, 109017, Russia.

where $V = S\,dx$ (S is a square of the section perpendicular to the direction of the flow). If $S = 1$, then

$$dm_i = \varphi_i C_i\, dx, \quad \text{or} \quad dm_i/dx = \varphi_i C_i.$$

Commonly, however, hydrothermal solution reacts with the rock and alters its chemical and mineral composition. Therefore between unaltered and totally altered rocks some transitional region occurs where concentration gradients initiate a range of consequent chemical reactions of dissolution and deposition of minerals. According to the local (or mozaic) equilibrium model of the metasomatic zonation by KORZHINSKII (1970), several mineral zones appear in this transitional region. The neighbor zones differ from each other by one mineral and are divided by sharp boundaries (replacing fronts) where metasomatic reactions are localized. Inside the zones equilibria are attained, and reaction rates are equal to zero. The rate of movement of the arbitrary section with constants C_i and q_i according to this model is

$$v = \varphi_i(\partial C_i/\partial q_i),$$

and the rate of the front (boundary between zones) movement is

$$v = \varphi_i(\Delta C_i/\Delta q_i)$$

where q_i is a content of component i in the rock. The local equilibrium model is applicable to the commonly described "monodirected" or vectoral zonation. The periodic or rhythmically banded zonation, in contrast, cannot be interpreted in terms of this model. Rhythmically banded structures, however, are rather common in various products of hydrothermal activity. There are known rhythmically banded garnet-diopside, spinel-forsterite, spinel-diopside and hedenbergite-wollastonite scarns. The banded structure occurs in one or two zones within the whole zoned scarn body. Quartz-adular rhythmical bands are common in upper (the most productive) parts of some epithermal gold-bearing veins, etc. Such structures previously localized near zone boundaries and inside zones of metasomatic zonation, are included in the total zonation pattern as its element.

The formation of such structures within individual zones is a feature of strong disequilibrium there are not inscribed in the mozaic equilibrium model. Some mechanisms of banded structure 11 formation in geological objects were studied by ORTOLEVA (1994). He proved that many of them are the results of self-organization in geochemical systems. One of the common mechanisms is the supersaturation-nucleation-solution cycle in the Ostwald-Liesegang pattern formation. The others are redox front and trapping front propagation. Models of chemical reaction front formation in the solution flow are worked out in ORTOLEVA et al. (1986). These models discussed front formation and its movement, as well as the possibility of temporal oscillations near the front (the cycle: supersaturation-nucleation-solution) due to redox reactions. Pyrite-goethite banding was studied as an example. In our paper we want to propose the model of metasomatic zonation with concentration waves and alternating mineral bands inside zones, far from their boundaries (far from reaction fronts). Previously we studied rhythmically banded wollastonite-hedenbergite scarns (RUSINOV et al., 1994) and made the model of its formation due to spatial dissipative structure in filtrating solution. The entire scarn body consists of consequent zones (from granite to limestone): garnet + wollastonite; garnet + diopside; wollastonite + subordinate hedenbergite. Only the last zone (the width of the zone is around 300–400 m) has rhythmically banded structure (Fig. 1a). The basic equations for the model describe reactions in the system $W + G + X + Y + Z + SiO_2$, where W is wollastonite, G—hedenbergite, $X - Fe^{2+}$, $Y - Fe^{3+}$, $Z - Ca^{2+}$. Silica is in excess and therefore is excluded from equations:

$$G \leftrightarrow X + Z; \quad X \leftrightarrow Y; \quad X + 2Y \rightarrow 3Y; \quad Z + 2Y \rightarrow X + Y + Z;$$

$$X + 2Y + Z \leftrightarrow G + 2Y; \quad Z + 2Y \leftrightarrow G + Y; \quad G \rightarrow Y + Z;$$

$$Z + 2Y \leftrightarrow W + 2Y; \quad \text{and} \quad Z \leftrightarrow W.$$

The study of the dynamics of this system demonstrates that under specific conditions there is a possibility of stable oscillations along the limit cycle (Fig. 1b), and numerical simulations show how the dissipative structure in solution and corresponding mineral bands evolve (Fig. 1c).

Now we propose a more generalized approach to the interpretation of metasomatic zonation, using the theory of travelling fronts and waves. The evolution of the metasomatic zonation can be discussed as a movement of one, two, or several reaction fronts with constant (but various for different fronts) velocities in the direction of solution filtration (inside individual zones enclosed between two fronts, distribution of the component concentrations is homogeneous). If this appears to be valid then the situation corresponds to travelling fronts regime. If the distribution of components inside the zone is heterogeneous and shows some extrema of concentration values, although the concentration waves still move with constant velocities, so it may be interpreted as travelling waves regime (ORTOLEVA et al., 1985).

Metasomatic Zonation in Terms of Travelling Fronts

The dynamics of evolution of metasomatic zonation in common form can be referred to as the equation of mass balance in the volume of the rock

$$\partial Q_i / \partial t + \partial I_i / \partial x = - F_i,$$

where $Q_i = q_i + pC_i$ is the total concentration of i in the rock (q_i) and in the solution (p is a porosity of rock), t—time, $I_i = uC_i - D_i(\partial C_i / \partial x)$ is a density of the

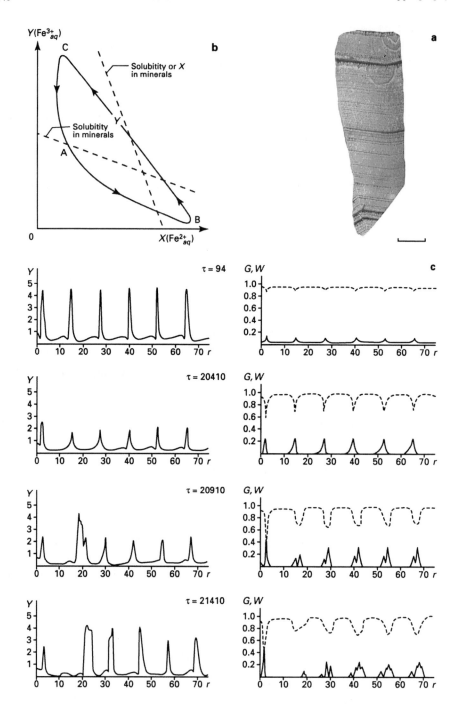

component i flux (including filtration as diffusion components), u—rate of filtration, D_i—diffusion coefficient, F_i—term describing the chemical kinetics of the redox reactions. In the diffusion operated processes the kinetic term is $F = 0$ and

$$\partial Q_i/\partial t + \partial I_i/\partial x = 0. \tag{1a}$$

The total pore volume in the rock is negligible compared with the rock volume and therefore we may approximate $Q = q$, and

$$\partial q_i/\partial t + \partial I_i/\partial x = 0. \tag{1b}$$

Introducing moving coordinates $r_i = x - v_i t$, we change the equations (1a,b) for

$$\partial Q_i/\partial t - v_i(\partial Q_i/\partial r_i) + \partial I_i/\partial r_i = 0, \tag{2a}$$

$$\partial q_i/\partial t - v_i(\partial q_i/\partial r_i) + \partial I_i/\partial r_i = 0. \tag{2b}$$

The total i flux is $\theta_i = I - v_i Q_i = I_i - v_i q_i$, and introducing the term w_{ij} which describes the kinetics of mineral j deposition or solution reactions, we have

$$\partial q_i/\partial t - v_i(\partial q_i/\partial r_i) - w_{ij} = 0, \tag{3a}$$

$$\partial I_i/\partial r_i + w_{ij} = 0. \tag{3b}$$

The kinetic term w_{ij} depends on two factors: kinetics of nucleation and kinetics of mineral growth. Solutions of the equations (2a,b) and (3a,b) with initial conditions $Q_i(0, r) = Q_i^0(r)$ and $q_i(0, r) = q_i^0$ are

$$Q_i(t, r) = Q_i^0(\xi) - (1/v_i) \int_{r_i}^{\infty} \partial[I_i(\tau, \eta)/\partial \eta]\, d\eta, \tag{4}$$

$$q_i(t, r) = q_i^0(\xi) + (1/v_i) \int_{r_i}^{\infty} w_{ij}(\tau, \eta)\, d\eta, \tag{5}$$

where $Q_i^0(\xi)$, $q_i^0(\xi)$, $I_i(\tau, \eta)$, $w_{ij}(\tau, \eta)$ are functions of variables $\xi = r_i + v_i t$, $\tau = t + (r_i - \eta)/v_i$, and η.

For $t \to \infty$ we have from (4) and (5) to (6) and (7a) correspondingly:

$$Q_i(\infty, r) = Q_i^0(\infty) - (1/v_i) \int_{r_i}^{\infty} [\partial I_i(\infty, \eta)/\partial \eta]\, d\eta = Q_i^0(\infty) - [I_i(\infty, \infty) - I_i(\infty, r)]/v_i \tag{6}$$

Figure 1
Rhythmically banded wollasonite-hedenbergite scarns (RUSINOV et al., 1994); (a)—photo of the sample, dark bands—hendenbergite, light—wollastonite, circle-like structure—datolite, scale-bar—1 cm; (b)—oscillations along limit cycle that form dissipative structure in solution and alternating bands in the scarn; (c)—results of the numerical simulations of the wollastonite-hedenbergite alternating bands (left part—evolution of the dissipative structure in solution at consequent time moments, right part—formation of hedenbergite bands in wollastonite matrix at the same moments), τ, r—nondimensional time and space correspondingly.

$$q_i(\infty, r) = q_i^0(\infty) + (1/v_i) \int_{r_i}^{\infty} w_{ij}(\infty, \eta) \, d\eta. \tag{7a}$$

$q_i^0 \gg p^0 C_i^0 - pC_i$, then

$$q_i = q_i^0 - [I_i(\infty, \infty) - I_i(\infty, r)]/v_i. \tag{7b}$$

The equilibrium (7b) allows estimation of the velocity of the front movement by substitution of q_i and I_i values, if $r_i = -\infty$ ($q_i = q_{io}$; $I_i = uC_{io}$; $I_i(\infty, \infty)uC_i^0$):

$$v_i = u(C_i^0 - C_{io})/(q_i^0 - q_{io}) = u\Delta C_i/q_i. \tag{8}$$

Here $q_i^0 - q_{io} = \Delta q_i = v_{ij} q_j^0$ (v_{ij}—stoichiometric coefficient of i in the mineral j; q_j^0—mineral j content in the zone distant from the front). The rate of component movement in the solution flow depends on the filtration coefficient which varies for different components. Therefore from (8) we have

$$v_i/u = \varphi_i \Delta C_i/\Delta q_i,$$

that adequately corresponds to the Korzhinskii's equation for the mozaic equilibrium model mentioned above. Consequently

$$v_i \Delta q_i/(\varphi_i \Delta C_i) = u = \text{const},$$

i.e., fronts are to move with constant velocities and the distribution of component concentrations remains stationary with elapsed time. Consequently the infiltration metasomatic zonation evolving in travelling fronts regime corresponds to the mozaic equilibrium model. It does not display concentration oscillations far from zone boundaries. All reactions between solution and minerals concentrate at replacement fronts.

Metasomatic Zonation as Travelling Fronts and Waves

We shall now discuss conditions under which the evolution of metasomatic infiltration zonation exhibits features of travelling waves. Replacement reactions are localized within the nearest vicinity of fronts (zone boundaries), therefore the distribution of the component around fronts is important for the model under discussion. From (3b), (7b) and taking in account that $I_i = v_i(q_i - q_i^0) + uC_i^0$ and $q_i^0 = q_{io} + u(C_i^0 - C_{io})/v_i$, we have

$$\left. \begin{array}{l} dC_i/dr_i = (u/D_i)C_i - (1/D_i)I_i \\ dI_i/dr_i = -w_{ij}(q_i, C_i) \end{array} \right\} \tag{9a}$$

$$\left. \begin{array}{l} dC_i/dr_i = (u/D_i)(C_i - C_i^0) - (v_i/D_i)(q_i - q_i^0) \\ dq_i/dr_i = -(1/v_i)w_{ij}(q_i, C_i) \end{array} \right\}. \tag{9b}$$

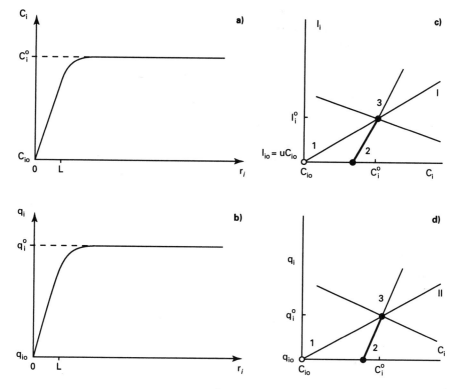

Figure 2
Distribution of the component i in the nearby zones and in the transition band between them (the i solubility C_i^0 is constant through the zone) for the travelling fronts model. a–b—spatial distribution of i in solution (a) and in solid phases (b); c–d—phase portraits of the system on the planes $(I_i - C_i)$ (c) and $(q_i - C_i)$ (d) with the saddle rest point. I_i is a flux without diffusion. Arrows show evolution paths from the back zone through the contact band to the rest point that characterizes the equilibrium inside the zone under study. Note that the path is to come to the centripetal separatris. Lines from the beginning of coordinate axes are: I—flux of i without diffusion $(I_i = uC_i)$; II—changing of i content in the solid phases $(q_i = q_{io} + (u/v_i)(C_i - C_{io}))$. Other solid lines are separatrisses. Numbered points correspond to composition of: 1—back zone; 2—boundary between zones; 3—rest point of the zone under study.

Systems of equations (9a,b) belong to autonomous systems describing free nonlinear oscillations. Therefore the trajectories on the phase plane (C_i, I_i) or (C_i, q_i) of solutions of these equations (Fig. 2) contain information concerning the distribution of the concentration of solutes and the content of the component in the rock. If the solubility (C_i^0) is constant inside the zone then rest points for these solutions correspond to the homogeneous state $(w_{ij} = 0)$ and are saddle ones unstable in the linear approximation. Hence all trajectories are unclosed and oscillations do not occur in systems under discussion.

In the natural systems the chemical reaction rates have finite values. Therefore the zone boundaries are not planes but transitional bands with measurable widths.

Their widths are equal to the distance at which solution attains equilibrium by component i with mineral j of the next zone:

$$L = u(C_i^0 - C_{io})/w_o, \qquad (10)$$

where w_o means the initial value of w_{ij}. Therefore the width of the transitional band depends on the filtration rate and kinetics of deposition-solution reactions of the mineral j: the band is narrow and the corresponding boundary is sharp if filtration is rather slow and/or reaction rate is high. Conversely, quick filtration and slow reactions make the transition gradual and transitional band wide. Far enough from the boundary inside the zone solution attains equilibrium with rocks and thus "mozaic equilibrium" occurs.

All the above discussed concern the case of constant solubility of i within the zone. If the solubility varies then some points with an extremely low and high content of mineral j can appear inside the zone due to varying C_i (C_i can exceed the solubility value around one or more points). This case corresponds to the travelling waves regime, and the "mozaic equilibrium" cannot be attained in such a system.

Metasomatic Zonation as Travelling Waves with Redox Reactions

If the component i can change its valency, accordingly the metasomatic replacement of the rock at the reaction fronts includes the following kinds of reactions:

$$Y + M_j \leftrightarrow X + Y \qquad (11a)$$

$$X \leftrightarrow Y + n\,e^- \qquad (11b)$$

$$X + Y \leftrightarrow M_k + X \qquad (11c)$$

where X—reduced and Y—oxidized forms of any component i (f.i., iron); M_j—mineral with reduced form, and M_k—mineral with oxidized forms of the same component. Subsequently the system of equations describing spatial distribution of concentration values and inputs of X and Y in the solution, and of minerals j and k in rocks for these reactions is as follows

$$\left. \begin{aligned}
dC_x/dr &= (u/D_x)C_y - (1/D_x)I_x, \\
dC_y/dr &= (u/D_y)C_y - (1/D_y)I_y, \\
dI_x/dr &= -p[(q'_x)^m w_j(C_x, C_y) + F_{xy}(C_x, C_y)], \\
dI_y/dr &= -p[(q'_y)^m w_k(C_x, C_y) + F_{xy}(C_x, C_y)], \\
dq'_x/dr &= -(p/v)(q'_x)^m w_j(C_x, C_y), \\
dq'_y/dr &= -(p/v)(q'_y)^m w_k(C_x, C_y),
\end{aligned} \right\} \qquad (12)$$

where $q'_{x(y)} = q_{x(y)} - q_{xo(yo)} = x_{j(yj)}q_{j(k)}$ is the concentration of X (Y) referred to the mineral $j(k)$; $(q'_{x(y)})^m w_{j(k)}(C_x, C_y) = w_{xj(yk)}$ is a kinetics of deposition (solution) of minerals j or k correspondingly; $m = 2/3$ or 1; $F_{xy}(C_x, C_y)$ is a general kinetics of the redox reaction; p is a porosity of the rock. The possibility of oscillating solutions of the equations (12) depends on the character of the trajectories in the phase space $(C_x, C_y, I_x, I_y, q'_x, q'_y)$, but the analysis can be simplified by discussion of only projections of the trajectories on the planes (C_x, C_y) and (I_x, I_y).

The stationary state of the system (12) is defined by equations

$$I_x = uC_x; \quad I_y = uC_y, \tag{13}$$

$$w_j(C_x, C_y) = 0; \quad w_k(C_x, C_y) = 0; \quad F_{xy}(C_x, C_y) = 0. \tag{14a}$$

So the rest point corresponds to

$$C_x = C_{xo}; \quad C_y = C_{yo}. \tag{14b}$$

and both of these values are to be solutions for three equations simultaneously:

$$w_j = 0; \quad w_k = 0; \quad F_{xy} = 0.$$

Such a unique coincidence is rather impossible in the natural processes. Therefore the system (12) has two rest points if $q'_x = 0$; $q'_y \to 0$; or $q'_x \to 0$ and $q'_y = 0$. If $q'_x > 0$ and $q'_y > 0$, the system has no rest points, and thermodynamic equilibrium cannot be attained. Oscillations in (12) are possible if both of the rest points are unstable. This means that reactions are to increase to complete dissolution of one of the minerals (j or k), and in the state of rest the content of one of the minerals is equal to zero. Thus two components related to the various valency of the one chemical element cannot be inert together (cannot form paragenetic assemblage of the two minerals because one of them must be dissolved). The analysis of the modified systems (15a,b) present different results.

$$\left.\begin{aligned}
dC_x/dr &= (u/D_x)C_x - (1/D_x)I_x \\
dC_y/dr &= (u/D_y)C_y - (1/D_y)I_y \\
dI_x/dr &= -p[(q'_x)^m w_j + F_{xy}] \\
dI_y/dr &= pF_{xy} \\
dq'_x/dr &= -(p/v)(q'_x)^m w_j
\end{aligned}\right\} \tag{15a}$$

$$\left.\begin{aligned}
dC_x/dr &= (u/D_x)C_x - (1/D_x)I_x \\
dC_y/dr &= (u/D_y)C_y - (1/D_y)I_y \\
dI_x/dr &= -pF_{xy} \\
dI_y/dr &= -p[(q'_y)^m w_k - F_{xy}] \\
dq'_y/dr &= -(p/v)(q'_y)^m w_k
\end{aligned}\right\}. \tag{15b}$$

Both of the systems (15a) and (15b) have at least one rest point opposite to the system (12). The rest points correspond to the state of thermodynamic equilibrium:

$$C_x = C_x^0; \quad C_y = C_y^0; \quad I_x = uC_x^0; \quad I_y = uC_y^0; \\ q'_{x(y)} = q_i^0 = u(C_x^0 + C_y^0 - C_{xo} - C_{yo})/v, \quad (16)$$

where $C_{x(y)}^0$ corresponds to the solubility of the mineral $j(k)$. The rest point coordinates are the solutions of the following equations:

$$(q'_{x(y)})^m w_{j(k)} + (-)F_{xy} = 0; \quad F_{xy} = 0.$$

Solution trajectories at the phase plane (C_x, C_y) for $I_x = uC_x$; $I_y = uC_y$ (diffusion is negligible) are defined with the following equation systems:

$$dC_x/dr = -(p/u)[(q'_x)^m w_j(C_x, C_y) + F_{xy}(C_x, C_y)], \\ dC_y/dr = (p/u)F_{xy}(C_x, C_y), \\ dq'_x/dr = -(p/v)(q_x)^m w_k, \quad (17a)$$

$$dC_x/dr = -(p/u)F_{xy}(C_x, C_y), \\ dC_y/dr = -(p/u)[(q'_y)^m w_k(C_x, C_y) - F_{xy}(C_x, C_y)], \\ dq'_y/dr = -(p/v)(q'_y)^m w_k, \quad (17b)$$

where $q'_{x(y)} > 0$ and $p > 0$.

The systems (17a,b) can be reduced by substitutions to

$$dC_x/dr = -[P_j(C_x, C_y)/u][Q(C_x, C_y)^m w_j(C_x, C_y) + F_{xy}(C_x, C_y)] \\ dC_y/dr = [P_j(C_x, C_y)/u]F_{xy}(C_x, C_y), \quad (18a)$$

$$dC_x/dr = -[P_k(C_x, C_y)/u]F_{xy}(C_x, C_y), \\ dC_y/dr = -[P_k(C_x, C_y)/u][Q(C_x, C_y)^m w_k(C_x, C_y) - F_{xy}(C_x, C_y)]. \quad (18b)$$

The rest points of the systems (18a,b) are defined like rest points of (15) as well as are phase portraits in the vicinity of rest points. Solution trajectories for the systems (18a,b) are closed, therefore some oscillating solutions of the system (15) are possible. Necessary conditions for such oscillating solutions are: 1) the rest point is not a saddle; 2) the rest point is unstable in the linear approximation; and 3) trajectories of solutions are enclosed in the closed contour encircling the rest point. The instability of rest points for (18a,b) depends on q_i^0 values.

Systems (18a,b) linearized in the vicinity of the rest points and represented as vector system are as follows:

$$dx/dr = \mathbf{M}x,$$

where

$$\mathbf{x} = \begin{Bmatrix} C_x - C_x^0 \\ C_y - C_y^0 \end{Bmatrix}; \quad \mathbf{M} = \begin{Bmatrix} -a - \alpha_j(q_i^0)^m; & -b - \beta_j(q_i^0)^m; \\ a - \alpha_k(q_i^0)^m; & b - \beta_k(q_i^0)^m; \end{Bmatrix}.$$

The coefficients here are: for the system (18a)—

$$\alpha_j = [P_j(C_x^0, C_y^0)/u](\partial w_j/\partial C_x); \quad \beta_j = [P_j(C_x^0, C_y^0)/u](\partial w_j/\partial C_y); \quad \alpha_k = 0; \quad \beta_k = 0;$$

$$a = [P_j(C_x^0, C_y^0)/u](\partial F_{xy}/\partial C_x); \quad b = [P_j(C_x^0, C_y^0)/u](\partial F_{xy}/\partial C_y);$$

and for the system (18b)—

$$\alpha_j = 0; \quad \beta_j = 0; \quad \alpha_k = [P_k(C_x^0, C_y^0)/u](\partial w_k/\partial C_x); \quad \beta_k = [P_k(C_x^0, C_y^0)/u](\partial w_k/\partial C_y);$$

$$a = [P_k(C_x^0, C_y^0)/u](\partial F_{xy}/\partial C_x); \quad b = [P_k(C_x^0, C_y^0)/u](\partial F_{xy}/\partial C_y).$$

For all derivatives $C_x = C_x^0$, and $C_y = C_y^0$.

Oscillatory instability of rest points (18a,b) is possible if the real part of the eigenvalues of the corresponding matrix $\mathbf{M}$ values (λ_1, λ_2) are positive. The λ_1 and λ_2 are roots of characteristic equation

$$\lambda^2 - (\alpha_j, \text{ or } \beta_k)(q_c - q)\lambda + Dq = 0,$$

so the necessary condition of oscillations in (18a,b) can be written as the system (19):

$$1 < b/a < \beta_j/\alpha_j, \tag{19a}$$

$$1 < b/a < \beta_k/\alpha_k, \tag{19b}$$

where (19a) and (19b) relate to the (18a) and (18b) correspondingly. If the right part of one of the inequalities (19a or b) is broken, then the equation (18a or b) becomes unstable, although no oscillations occur. If both inequalities (19) are broken, oscillations can appear in the initial system (12), because both of its rest points are unstable.

More precise conditions of oscillations in (18a,b) can be derived from the characteristic equation listed above. It can be rewritten as

$$\lambda^2 - (\alpha_j \text{ or } \beta_k)(q_c - q)\lambda + Dq = 0,$$

where $q = (q_i^0)^m$; $q_c = (b-a)/(\alpha_j \text{ or } \beta_k)$; $D = a(\beta_j \text{ or } \beta_k) - b(\alpha_j \text{ or } \alpha_k)$.

It means that behavior of the system (18) around the rest point depends on q value. This dependence is shown in Figure 3. The value $q = q_c$ is the bifurcation value. The rest point is stable if $q > q_c$ and unstable if $q < q_c$. If decreasing q attains q_c value ($q = q_c$) then stable limit cycle can appear in (18a,b). Its period corresponds to the Hopf theorem $T \cong 2\pi/\sqrt{Dq_c}$. If solutions of (18a,b) are limited then stable limit cycles enclose rest points of the systems in III and IV (see Fig. 3). The conditions limiting the solutions (18a,b) are derived from the phase portraits of the systems on the plane (C_x, C_y)

$$dC_x/dC_y = -1 - w_x/F_{xy}, \tag{20a}$$

$$dC_y/dC_x = -1 + w_y/F_{xy}. \qquad (20b)$$

The equation (20a) describes the system (18a), and equation (20b) describes the system (18b). Here $w_x = Q(C_x, C_y)^m w_j(C_x, C_y)$; $w_y = Q(C_x, C_y)^m w_k(C_x, C_y)$; F_{xy} is a function of C_x, C_y; $Q \geq 0$; w_j, w_k, and F_{xy} can be as positive as negative. The following relations are considered valid as well:

$$F_{xy}(0, C_y) < 0; \quad w_j(0, C_y) < 0; \quad F_{xy}(C_x, 0) > 0; \quad w_k(C_x, 0) < 0; \quad F_{xy}(0, 0) = 0. \qquad (21a)$$

The directions of solution trajectories correspond to the three lines (Fig. 4) 1) $C_x^f = f(C_y)$ or $C_y^f = f(C_x)$; 2) $C_x^g = g(C_y)$ or $C_y^g = g(C_x)$; 3) $C_x^w = w(C_y)$ or $C_y^w = w(C_x)$, if the following conditions are valid

$$F_{xy}(C_x^f, C_y) = 0; \quad F_{xy}(C_x, C_y^f) = 0; \quad F_{xy}(C_x^g, C_y) + w_j(C_x^g, C_y)Q(C_x^g, C_y)^m = 0;$$

$$F_{xy}(C_x, C_y^g) - w_k(C_x, C_y^g)Q(C_x, C_y^g)^m = 0; \quad w_j(C_x^w, C_y) = 0; \quad w_k(C_x, C_y^w) = 0.$$

Also necessary are relations (21b)

$$\left. \begin{array}{l} w(0) > g(0) > f(0) = 0; \quad \partial f/\partial C_i|_{C_i=0} > 0; \quad \partial g/\partial C_i|_{C_i=0} > 0 \\ f(C_y^0) = g(C_y^0) = w(C_y^0) = C_x^0; \quad f(C_x^0) = g(C_x^0) = w(C_x^0) = C_y^0 \end{array} \right\} \qquad (21b)$$

where $i = X, Y$.

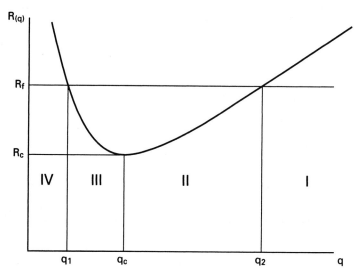

Figure 3
Behavior of the system related to the q value around rest points. Here $R = q + q_c^2 q$; $R_c = 2q_c$; $R_f = R_c + 4D/(\alpha_f^2$ or $\beta_k^2)$; $q_{1,2} = (R_f \pm \sqrt{R_f^2 - R_c^2})/2$. I–IV—fields with various behavior of the system: I—$q > q_2$ and $R > R_f$—stable knot, the system tends monotonously to equilibrium; II—$q_c < q < q_2$ and $R_c < R < R_f$—stable focus, the system tends to equilibrium through attenuating oscillations; III—$q_1 < q < q_c$ and $R_c < R < R_f$—unstable focus, the system disposes itself of equilibrium oscillating with increasing amplitude; IV—$q < q_1$ and $R > R_f$—unstable knot, the system disposes itself of equilibrium monotonously.

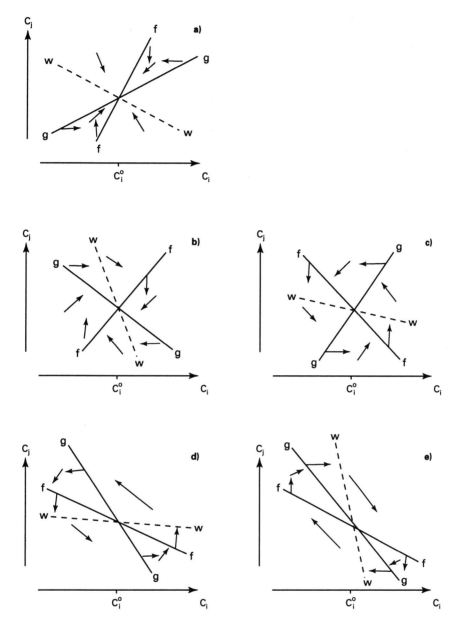

Figure 4

Phase portraits of solutions of the equilibria (18a,b) on the plane (C_j, C_i) showing the behavior of trajectories around the equilibrium state. Arrows show the shift of the trajectories to their attractors. Lines f, g, and w correspond to $C_x(y)$ values under following conditions: $F_{xy} = 0$ (f); $F_{xy} + w_x = 0$ (g); $w_x = 0$ (w); F_{xy}—redox reaction kinetics; $w_{x(y)}$—mass exchange (of X and Y) reaction kinetics between solution and solids. a—stable knot corresponding to the vector zonation; b, c—stable focus corresponding to travelling waves near back zone boundary (local rhythmical banding); d, e—unstable focus evolving into limit cycle corresponding to travelling waves and rhythmical banding throughout the zone.

Under conditions (21a,b) trajectories of solutions (18a,b) lie on the phase plane (C_x, C_y) in the region of positive C_x and C_y (Fig. 2). The behavior of solutions (18a,b) depends on functions f, g, w in the vicinity of the rest point (Fig. 3). The trajectories become closed and encircle the unstable rest point only under special conditions:

$$\partial f/\partial C_i|_{C_i=C_i^0} < 0; \quad \partial g/\partial C_i|_{C_i=C_i^0} < 0 \quad (i = X, Y). \tag{21c}$$

In terms of (19a,b) these conditions can be written as

$$b > 0; \quad a > 0; \quad -b/a < 0; \quad -[b + \beta_j(q_i^0)^m]/[a + \alpha_j(q_i^0)^m] < 0$$

or

$$-[a - \alpha_k(q_i^0)^m]/[b - \alpha_k(q_i^0)^m] < 0.$$

Lines f and g on the phase plane cannot cross the coordinate axes, which means

$$C_i \to \infty; \quad f(C_i) \to \text{const}; \quad g(C_i) \to \text{const}; \quad w(C_i) \to \text{const}; \quad \partial f/\partial C_i \to 0;$$

$$\partial g/\partial C_i \to 0 \quad (i = X, Y). \tag{21d}$$

If all these conditions are valid, then the trajectories of solutions (18a,b) necessarily cross lines f and g (Fig. 4), change their direction in these points and therefore cannot cross-coordinate axes. So they become closed orbits. In the simplest examples such trajectories are limit cycles. According to the relations (21a–d), lines f and g pass through the maximum point before attaining the rest point, and then they asymptotically attain the constant value if the argument grows. Such behavior appears to be possible if the redox kinetics (F_{xy}) contains at least one three-molecular stage, which denotes the occurrence of the autocatalitic stage in the general reaction kinetics.

RUSINOV and ZHUKOV (1994) described the mechanisms of influence of redox reaction

$$Fe^{2+} \to Fe^{3+} + e^-$$

on the formation of the rhythmically banded zonation in bauxites and wollastonite-hedenbergite scarns. Occurrence of two- and three-core complexes with Fe-ions and their activity in redox processes in the solutions favors the appearance of the autocatalytic three-molecular stage of reaction.

Figure 5 presents some of the simplest examples of the formation of limit cycles under boundary conditions corresponding to metasomatic systems. The occurrence of some complexes of various rock-forming elements with ferrous and ferric ions (Na, Ca, and Mg with Fe^{3+}, but Al with Fe^{2+}) can involve most petrogenic components and rock-forming minerals into the oscillations and formation of rhythmically banded structures in metasomatic rocks.

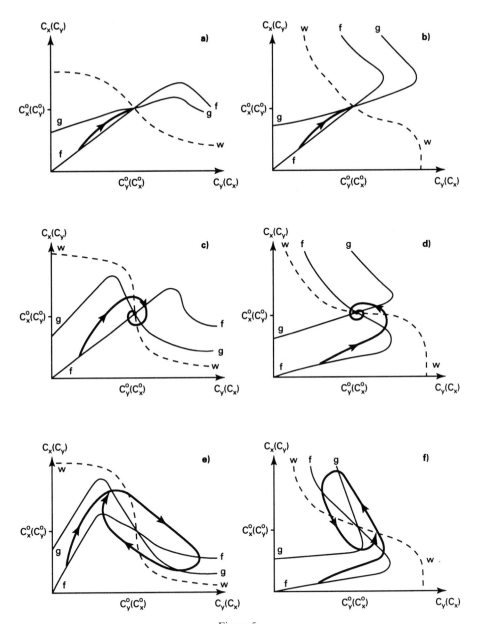

Figure 5
Phase portraits of the solution of the equilibria (18a,b) on the phase plane (C_x, C_y) showing various types of trajectories (thick lines). a, b—stable knot (travelling fronts) cause vectoral zonation; c, d—stable focus (decaying travelling waves) causes local rhythmical banding near the back front of the zone; e, f—unstable focus inside the limited cycle (travelling waves) causes rhythmically banded zonation. Symbols and explanations are the same as for Figure 3.

If one of the equations (18a,b) is broken accordingly the metasomatic zonation develops like a system of travelling fronts with mozaic equilibrium inside the zones. Thus this version of the process corresponds to the case of travelling fronts and to the model of local (mozaic) equilibrium. The common vectoral zonation pattern appears in this case, but it can be complicated with locally occurring decaying oscillations. Such oscillations are located in the vicinity of moving fronts where rhythmical banding can appear. Oscillations commonly preceded the attainment of mozaic equilibrium and of homogeneous components' distributions inside the individual zone.

Acknowledgements

The authors are grateful to the reviewer for very careful reading of the manuscript, its discussion, and helpful remarks.

This work has financial support of the Russian Foundation for Basic Research, Project no. 97–05–64131.

REFERENCES

KORZHINSKII, D. S., *Theory of Metasomatic Zoning* (Oxford, Clarendon Press 1970).

ORTOLEVA, P., AUCHMUTY, G., CHADAM, J., HETTMER, J., MERINO, E., MOORE, C. H., and RIPLEY, E. (1986), *Redox Propagation and Banding Modelities*, Physica D *19* (3), 334–353.

ORTOLEVA, P., *Geochemical Selforganization* (Oxford, Oxford University Press, Clarendon Press 1994).

ORTOLEVA, P., SCHMIDT, S. L., and STEVEN, L., *The structure and variety of chemical waves*. In *Oscillations and Travelling Waves in Chemical Systems* (Field, R. J., and Burger, M., eds.) (New York–Chichester–Brisbane, John Wiley and Sons 1985) pp. 333–418.

RUSINOV, V. L., KUDRYA, P. F., LAPUTINA, I. P., and KUZMINA, O. V. (1994), *Periodical Metasomatic Zonation in Pyroxene-Wollastonite Scarns*, Petrology *2* (6), 510–525.

RUSINOV, V. L., and ZHUKOV, V. V. (1994), *A Genetic Model for Rhythmically Banded Structure in Exogene and Hydrothermal-Metasomatic Systems*, Geol. of Ore Deposits *36* (6), 529–535.

(Received March 31, 1998, revised May 11, 1999, accepted June 1, 1999)

To access this journal online:
http://www.birkhauser.ch

Wavelet Analysis of Nonstationary and Chaotic Time Series with an Application to the Climate Change Problem

D. M. Sonechkin[1] and N. M. Datsenko[1]

Abstract—Our aim here is to show merits of the so-called wavelet transform of time series as an analyzing tool of the simultaneously nonstationary and chaotic processes that usually appear in the Earth sciences. The merits are illustrated on some surrogate time series as well as the natural ones that are observed in the present-day climate change problem. As a result of a wavelet transform of the natural hemispheric averaged surface air temperature time series a crossover scale is recognized that cuts off the intra- and interdecadal temperature oscillations, which are obviously internally induced, from a secular trend-like component of unknown origin.

Key words: Wavelets, scaling, multifractality, climate dynamics and change.

1. Introduction

It is generally accepted in the Earth sciences to analyze observational data time series by means of such traditional techniques as the Fourier transform, spectral and regression analyses. All of these techniques are based on an implicit assumption of stationarity of processes of interest. But, the assumption is in contradiction with the property of a majority of the real processes to be chaotic and, at the same time, essentially nonstationary in the overall length for most observed time series.

However, it is admissible to believe in many cases that such observed nonstationarity is artificial. It is induced by the insufficient length of a time series being analyzed. The illusion is supported partly by viewing the processes through the framework of the mathematical theory of the nonlinear dynamical systems. The so-called strange attractor (an attractive set of trajectories in a phase-space of the dynamic system of interest) is a generally accepted mathematical model of such behavior. As it is well known (ECKMANN and RUELLE, 1985), there is an invariant ergodic probability measure concentrated on each strange attractor. Therefore, a typical trajectory of such an attractor must be strictly staionary in the statistical sense. However, this is not a hindrance for each of such trajectories to show

---

[1] Hydrometeorological Research Centre of Russia, Bolshoy Predtechensky Lane 9/13, Moscow 123242, Russia, E-mail: dsonech@mskw.mecom.ru

variations over a very wide range of time scales. Some of these inherent variations can be so great and long-lived that, looking at a bit of such a trajectory, one can assume that the process being analyzed is truly nonstationary. For this reason, the nonstationarity can be an artifact of finiteness for every observed time series.

Nevertheless, the nonstationarity may be presumed to be real if a process of interest is governed by an evolutionary equation, as for example the well-known equation of the heat balance of the atmosphere

$$dT(t)/dt = R_{in}(t) - R_{out}(t) = p(t) \qquad (1)$$

where $R_{in}(t)$ and $R_{out}(t)$ are the incoming and outgoing radiations, respectively. The impulses $p(t)$ are apparent random instaneous radiational imbalances of the climate system. The interannual differences $\Delta T(t) = T(t) - T(t-1)$ imitate the imbalances. The energy-balance model based on (1) is widely used in meteorology after the BUDYKO pioneering work (1969).

From the mathematical point of view (see MANDELBROT, 1982), Equation (1) depicts trajectories of the so-called Brownian motion (**Bm**) of a particle. A prominent feature of **Bm** is its general nonstationarity. The case of the anthropogenic origin of the present-day climate warming, for example, may be treated (SONECHKIN et al., 1997) as **Bm** with the expectation value of $p(t)$ being non-zero. But, even if the expectation value were equal to zero, and the probability distribution of $p(t)$ were invariant in time, Equation (1) would reveal a nonstationary process with stationary increments. It must be taken in mind, however, that the imbalances of the climate system are undoubtedly dependent on the current state of the system. Therefore one can presume that the Langevin form of the heat balance equation

$$dT(t)/dt = R_{in}(t) - R_{out}(t) - \gamma(T(t) - T^*)) \qquad (2)$$

is more appropriate as a model of the real climate dynamics. The third term on the right side of (2) depicts explicitly a continuous tendency of the system to gain equilibrium. In case such tendency is strong (γ very large), nonstationarity of the resulting **Bm**-process can be damped.

The nonstationarity of all trajectories in the integral scale of the overall length of each observed time series can be frequently combined with their chaotic (apparently random) oscillations over relatively short-time scales. Thus, an appropriate analysis must take into account both of the above peculiarities of the processes: their integral nonstationarity that manifests itself as a super low-frequency (otherwise speaking, a trend-like) behavior of the process, and their relative high-frequency chaoticity.

The momentary (sliding or "windowed") Fourier transform of time series was the first step towards such an analysis. A relatively new and more powerful tool of the analysis is the so-called wavelet transform (**WT**) of the time series (RIOUL and FLANDRIN, 1992).

In Section 2, we briefly cite the basic definitions of **WT** and depict its specific useful properties. In Section 3, the **WT** technique is applied to the problem of the present-day climate warming. **WT** patterns of the available hemispheric air temperature time series are demonstrated, and some features of these are indicated. In particular, an unique warming trend is revealed by means of subtractions of inverted **WT**s from the primary series. In order to attribute these features to a certain physical origin, in Section 4 **WT** patterns of different surrogate time series are considered which are formed by means of random shuffles of the data-points of the real temperature series. It became interesting to compare self-similar scaling properties of both types of the series as they are revealed by **WT**. Because the hemispheric series are rather short (about 150 data-points only), some more extended local air temperature series are analyzed in Section 5 in order to substantiate the conclusions of Sections 3 and 4 concerning the character of the current climate warming.

2. Wavelet Transforms of Time Series. Definition and Properties

The continuous **WT** of a function $T(t)$ is its convolution with a family of (generally complex-valued) $G((t-b)/a)$ functions (see GROSSMANN et al., 1989)

$$WT(b, a) = a^{-1/2} \int_{-\infty}^{\infty} T(t) G^*((t-b)/a) \, dt \tag{3}$$

where the asterisk means the complex conjugate. The members of the family are obtained from a single "mother" wavelet function $G(t)$ by means of a variety of its dilations and translations. The mother function would be chosen from the square-integrable (and therefore sign changeable) functions which are localized near the zero value of their argument, so that the mean value of the wavelet function and, possibly, some next moments are equal to zero

$$\int_{-\infty}^{\infty} G(t) t^z \, dt = 0, \quad z \le Z. \tag{4}$$

This so-called admissibility condition guarantees good localization of **WT** in the time and frequency domains simultaneously. It makes **WT** to be a proper tool for a "local" analysis of transient events of time series over a wide range of scales unlike the "global" Fourier transform. When the admissibility condition (4) with $Z = 1$ is chosen, the $WT(b, a)$-coefficients are insensitive to any linear trend of the series under transformation independently of any (artificial or internally inherent) origin of the trend. Hence the coefficients from scale-dependent sequences which are strictly stationary in time. The temporal mean values of these sequences are equal to zero, however their variances and other higher statistical moments are functions of the scale.

The continuous **WT** is reversible for the square-integrable functions

$$\hat{T}(t) = \int_{-\infty}^{\infty} \int_{-\infty}^{\infty} WT(b,a) G((t-b)/a) a^{-2} \, db \, da = T(t). \tag{5}$$

If we add the function being transformed by a constant function or by a function of constant growth then the **WT** of the function does not change in principle. Thus, if the time series being transformed is a sum of a square-integrable and a monotonically varying function, then the latter component of the sum will not be included within the reconstructed series (5). Below we will see that this property can partly become lost in practical **WT** evaluations of finite time series.

Owing to a similar shape of all members of a wavelet family, **WT** preserves the self-similarity scaling which is known as a dynamically inherent property of the deterministic chaos (ECKMANN and RUELLE, 1985):

$$\Delta T(\lambda \tau) \stackrel{d}{=} \lambda^{H(t)} \Delta T(\tau), \tag{6}$$

where $\Delta T(\tau) = T(t+\tau) - T(t)$. Note that (6) must be tractable in the sense of the equality of the probability distributions of the compared quantities.

Equation (6) is transformed into the equality

$$WT(b, \lambda a) \stackrel{d}{=} \lambda^{H(t)+0.5} WT(b, a). \tag{7}$$

As a result, the $H(t)$-spectrum can be estimated on the basis of the scale-dependent behavior of the **WT** coefficients. We will use a simple method for this purpose. It consists of calculations of the following generalized variances (for $q > 0$):

$$VAR(q, a) = \lim_{T_{MAX} \to \infty} T_{MAX}^{-1} \int_{T_{MAX}} |WT(b, a)|^a \, db = \text{const} * a^{\zeta(q)}. \tag{8}$$

Here T_{MAX}—the length of the series used, and

$$\zeta(q) = qH + \delta H(q), \tag{9}$$

where $H = \zeta(1)$ is the well-known Hurst exponent (MANDELBROT, 1982) and $\delta H(q)$ are multifractal corrections of the scaling. The corrections can be estimated on the base of the scale-dependent behavior of the ratios

$$VAR(q, a)/(VAR(1, a))^q = \text{const} * a^{\delta H(q)}. \tag{10}$$

Choosing an appropriate lattice of the values of the dilation and translation parameters, for example, generated as $a = a_0^m$, $b = nb_0 a_0^m$, $m = 1, 2, \ldots$, $n = 1, 2, \ldots$, $a_0 > 0$, $b_0 \neq 0$, we can distinguish an almost orthogonal (the so-called frame of **WT**, see DAUBECHIES, 1988) or truly orthogonal (see MEYER, 1989) set from a continuous family of the wavelet functions

$$WT(b, a) = a^{-1/2} \sum_t T(t) G^*((t-b)/a). \tag{11}$$

If a_0, b_0 are chosen very close to 1 and 0, respectively, then the resulting discrete **WT** will be close to the generating continuous **WT**, and therefore, it can be used for the purpose of a practical numerical reproduction of **WT** patterns based on the interpolation of a discrete set of the **WT** coefficients. Such a discrete **WT** admits fine-grained reconstruction of the primary series by the series

$$T(t) = \sum_{a \leq a_{max}} a^{-2} \sum_{b} WT(b, a) G((t-b)/a) + RES(t) \tag{12}$$

that takes into account only the scales less than a chosen maximal scale a_{max}. The residual term $RES(t)$ of (12) describes a low-pass part of the primary series. In particular, $RES(t)$ depicts a nonstationary trend-like component of the series being transformed if the series is a realization of a nonsquare-integrable function, and thus, a continuous trend exists indeed. Certainly, $RES(t)$ also contains the super low-frequency oscillations (of the scales behind the maximal scale) of the square-integrable series.

In practice, the accuracy of the corresponding calculations of (11), (12) is strongly influenced by the shape of the mother function. In calculations which follow, we use, as an example, the so-called Mexican hat wavelet function

$$G(t) = (2/\sqrt{3}) \pi^{-1/4} (1 - t^2) \exp(-t^2/2). \tag{13}$$

$Z = 1$ for the Mexican hat, i.e., this function is not sensitive to any constant background and any trend of the series under transformation. It is less oscillatory as compared to other well-known mother wavelet functions. For this season, this mother function is most handy for numerical evaluation of the **WT** coefficients of relatively short scales, which are commensurable with the data discreteness of time serves being transformed.

To this point we stress very much that **WT** patterns are very sensitive to finiteness of every observed data series. If the finiteness is not taken into account, some artificial jumps at both ends of the series rise. Indeed, for the translation parameter values, which are rather close to the ends of the series, we are confronted with the need to calculate products of the nonzero values of the wavelet functions with data values of the points, which are located outside of the series and actually absent. Neglect of these products in the convolution (3) means in fact that we assume these data-points to be zero and the ends to coincide with artificial stepwise jumps.

The jumps disturb the boundary parts of **WT** patterns. The disturbance usually consists of an artificial increase of modula of **WT** coefficients. The methods to remove the mean value of the series or the series cycling, which are standard procedures when calculating the traditional Fourier transform, are inadequate for **WT** of essentially nonstationary series. It is better (MEYERS et al., 1993) to supplement the ends of the series with additional lots of data. Various forms of such supplements can be chosen on the base of any available information.

Our idea of the procedure implies reduction of the boundary **WT** coefficients to their most likely values. The latter values can be estimated using the probability distribution of **WT** coefficients of the corresponding middle parts of the series. At this point we must recall that $WT(b, a)$-sequences must be statistically stationary. Therefore the reduction of some boundary **WT** coefficients to their most likely values seems to be well-founded.

As the first guess of the supplements, we recommend use of the normals of some of the earliest and latest parts of the series. In order to obtain the second guess, it is necessary to decompose the first guess **WT** pattern into an oscillatory and a trend-like component, and then extrapolate the latter component to the supplements. If the new boundary **WT** coefficients will not be in sufficient conformity with the most likely ones, the procedure must be repeated.

3. Wavelet Transforms of the Hemispheric Air Temperature Time Series

To date the hemispheric mean surface air temperature time series remains one of the basic data sources for monitoring the current global climate warming. The most credible of such series are those created and continuously improved and updated by a group of climatologists from the United Kingdom. Here we use the latest available version of their series (JONES et al., 1994) for the 1854–1993 record period. The series are represented in terms of annual mean anomalies of the global net of the surface meteorological observations. The anomalies are departured from the 1951–1980 normals, and then hemispheric averaged. In the consideration below we use as the first guess of the series' supplements the mean values of the anomalies for the earliest and latest twenty years of the records. The lengths of the supplements were chosen to be 50 years long.

Figure 1 shows both hemispheric series (marked by the letter "A" in the figure). One can see that the series display a general warming about 0.5°C over the period of the records. It is believed (HOUGHTON et al., 1996) that the "urban warming" contributes only 0.1°C to this general warming. Additionally the warming is aggravated by heavy warming-cooling transitions on a very wide range of time scales from interannual extending to interdecadal. Such a mixture of the multiscale variations of the temperatures makes the problem of the current climate warming detection and attribution exceedingly complex.

In Figure 1 the two-dimensional **WT** patterns for both hemispheric series (**NH**—for the Northern Hemisphere and **SH**—for the Southern Hemisphere) on the $(b, \ln a)$—half-plane are also shown. The **WT** patterns display a kind of ordering, i.e., some tree-like structures of branches of the positive and negative polarity domains of the **WT** coefficients upward to shorter scales. The lines of the **WT** local extrema form a skeleton of these branches. It is easily seen that each branch propagates across the scales towards a point of these series that corresponds to a sharp variation of temperature. Because many of the finest of these variations are influenced by different inaccuracies of the series, the tree-like structures are undoubtedly distorted to a certain extent by these inaccuracies.

The tree-like structures over larger scales seem to be more correct. For example, there exists an enormous negative polarity domain in the left bottom part of both hemispheric **WT** patterns (before the end of 1940s) and a positive polarity domain

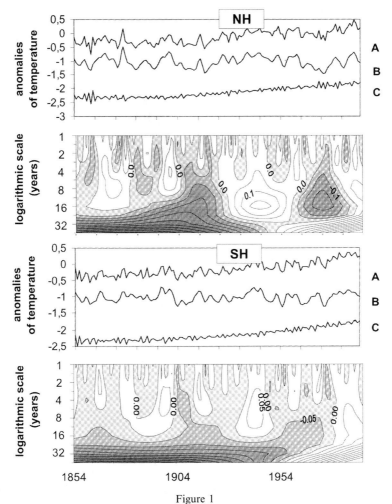

Figure 1

Wavelet transform of the extended (1854–1993) annual mean, hemispherically averaged (NH—the Northern Hemisphere, SH—the Southern Hemisphere) surface air temperature anomalies departured from the 1951–1980 normals (from JONES et al., 1994). Negative domains of the **WT** patterns are shaded. **A**—the temperature series being transformed, **B**—their stationary oscillatory components evaluated by the inverse wavelet transform (12) with the maximal scale $a_{max} = 22$ years, **C**—the trend-like components evaluated as residuals after subtraction of an oscillatory component from a primary time series. The **B**, **C** graphs have been vertically offset so as not to overlap.

on the right bottom half of these. As the scale decreases to about $a = 22$ years, at the right of the **WT** pattern of the Northern Hemisphere (within 1960–1980) a new negative polarity domain appears that takes the place of the underlying positive polarity domain almost completely. At the same scale, a new positive polarity domain (1930–1950) replaces a part of the underlying negative polarity domain. This branching corresponds to the well-known warming-cooling transition of the Northern hemispheric temperature during the middle of the 20th century. Despite a relative cooling is not seen explicitly in the Southern Hemisphere series during the second half of the 20th century (refer to this opinion in: SCHLESINGER and RAMANKUTTY, 1994), the shown **WT** pattern makes a visible signal of this transition there.

Furthermore, new branches appear towards the tops of both **WT** patterns. At the same time, both of the tree-like structures of Figure 1 start from the scale of about 22 years, and over the largest scales there is a strong, almost linear growth of the **WT** coefficients with time. Therefore, we have chosen the scale of 22 years as the crossover scale for decomposition of the series under analysis onto an oscillatory and a trend-like component.

The oscillatory components (marked by the letter "B" in Fig. 1) were evaluated by an inverse **WT** according to the formula (12). The above crossover scale was used as the maximal scale in this formula. It should be taken into account that the "effective length" of the Mexican hat wavelet function (13) is about 4 time more than one unit of the a-scale. Therefore, this crossover scale corresponds to the temporal cells of the series of approximately 80 years long, i.e., to half of the overall length of the series. It is evident that it is absurd to consider any cell of larger scale on the basis of the disposed data records.

The typical temperature variation due to extracted oscillatory components is about 0.5°C per a typical oscillatory period of about 10 years. By reason of strong stationarity of the oscillatory components, numerous traditional statistical tools, such as the regression analysis etc., can be used with a certain success in order to depict the oscillatory components and extrapolate those most forthcoming regarding the typical oscillatory period. In particular, on the basis of such extrapolation, we may conclude (SONECHKIN et al., 1996 and 1997) that the end of the 20th century will probably be cooler as compared with the observed years of the 1990s.

The trend-like components (marked by the letter "C" in Fig. 1) were evaluated as residuals of subtractions of the above oscillatory components ("B") from the primary time series ("A"). It must be mentioned that, by the reason of the finiteness of the "effective length" of the wavelet function and the discreteness of the primary series, it is impossible to evaluate the cells of the temperature series; the length of which is less than four years. Therefore the revealed oscillatory components contain no contribution of these shortest cells. Consequently, the shortest cells became included into the residuals. The cells manifest themselves in the residuals as a very high-frequency noise imposed upon truly super low-frequency (trend-like) variations of temperature.

Fortunately, the resulting signal/noise ratio of the residuals is so high that this high-frequency noise does not prevent recognition of the truly trend-like behavior of the hemispheric temperature time series, unlike such problem in the traditional statistical climate change detection studies (see: WOODWARD and GRAY, 1993). Using the known two-phase regression model (HINKLEY, 1971, see also: SOLOV, 1988) we estimated both extracted trend-like components to be equal to

$$T(t) = \text{const} + k^*x + E(t), \quad x = t - 1900, \quad t = 1900, \ldots, 1993$$

$$T(t) = \text{const} + E(t), \quad t < 1900 \tag{14}$$

where $k = 0.57°C$ and $0.58°C$ per 100 years for the Northern and Southern Hemispheres, respectively. Both hemispheric $E(t)$ form independent sequences of noise with mean zero and root-mean-square deviation of about $0.05°C$, evidencing the interhemispheric difference between the indicated trend increments to be insignificant.

4. Merits of WT as Illustrated for Some Surrogate Time Series

In order to be confident in the ability of **WT** to detect true trend-like behavior, we consider below **WT**s of several surrogate time series obtained by means of random shuffles of the terms of the real hemispheric temperature series.

Taking a random shuffle of these terms, we expect to obtain an undoubtedly random series such as a white noise with the same probability distribution that is inherent to the bare series. Certainly the above-depicted trend-like components of the bare series clearly must be destroyed as well as the tree-like structures of those series. Adding then such a strong stationary surrogate series by an artificial trend-like component, we can watch the response of the respective **WT** pattern to the trend-like component on the background of pure noises.

Figure 2 illustrates two examples of such pure random series (**A**) which are created by shuffles of the real temperatures of the Northern and Southern Hemispheres. Many cells of these surrogate series seem to be quite similar between hemispheric series. This fact indirectly evidences a strong interhemispheric coupling of the real temperature dynamics.

Yet more important, these **WT** patterns display tree-like structures similarly with the real series, because, even though consecutive terms of the surrogate series are not correlated, their **WT** patterns display homogeneous domains, the sizes of which increase with the scale. Furthermore, the **WT** patterns display an apparent crossover from their tree-like (oscillatory) structures to a trend-like behavior over the largest scales being considered (in the bottoms of the **WT** patterns in Fig. 2). Thus the phenomenon of this crossover must be interpreted as an artifact of finiteness of the series under transformation. Indeed, decompositions of both

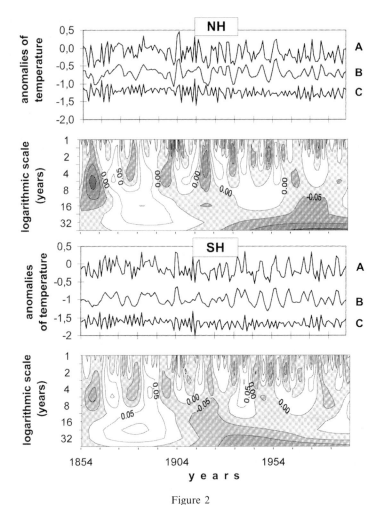

Figure 2
The same as Figure 1 but for the surrogate time series which were formed by random shuffles (synchronously for both hemispheres) of the members of the real temperature series from JONES et al. (1994).

surrogate series, based on this crossover scale as the maximal scale of the inverse **WT** by the formula (12), deflate the impression of the macroscale variations of temperatures in these surrogate series. Both trend-like components that resulted from these decompositions appear to be constant functions aggravated by stationary noises.

Figure 3 displays the same pure random series added by an artificial trend. The data-points of these series were firstly divided by two and then added to a linear trend with the increment $k = 0.6°C$ per 100 years. The goal was to create an essentially nonstationary and, simultaneously, pure random series, the general variability of which would be equal to the general variability of the real hemispheric series. Although the tree-like structures of these new surrogate series look quite similar to those presented in Figure 2, the bottoms of their **WT** patterns in Figure 3 display another crossover of the opposite sign, i.e., there is an impression of a macroscale growth of temperatures in these new series. This growth seems to be very similar to the corresponding growth in the real series (Fig. 1). Moreover, the whole **WT** patterns in Figure 3 appear quite similar to the whole **WT** patterns of the real series. The decomposition of the **WT** patterns in Figure 2 produce the trend-like components that are practically indistinguishable from such components of the real series.

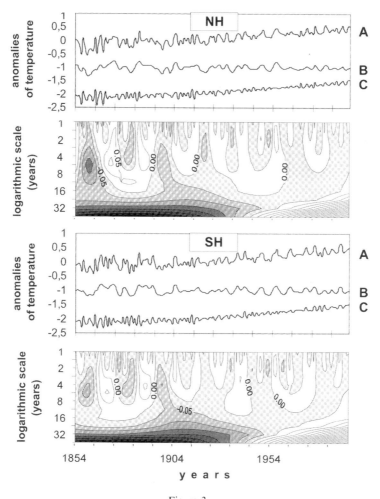

Figure 3
The same as Figure 2, but each shuffled member was divided by two after which the series was added by the linear trend with the increment of 0.6°C per 100 years.

Table 1

Scaling exponents of the real (A), pure random without any trend (B) and with an imposed trend (C), Brownian motion without any trend (D) and with an imposed trend (E) for the Northern (NH) and Southern (SH) hemispheric temperature time series

	A		B		C		D		E	
	NH	SH	NH	SH	NH	SH	NH	SH	NH	SH
$\zeta(1)$	0.08	−0.19	−0.58	−0.68	−0.42	−0.60	1.10	0.58	1.10	0.14
$\zeta(2)$	0.11	−0.39	−1.17	−1.35	−0.77	1.07	2.21	1.11	2.21	0.96
$\zeta(3)$	0.10	−0.63	−1.78	−2.05	−1.02	−1.53	3.25	1.59	3.26	1.61
$\zeta(4)$	0.01	−0.92	−2.39	−2.78	−1.31	−1.98	4.29	2.03	4.27	2.09

It would be premature, however, to conclude that the nature of the actual climate warming is the same as it is in these new surrogate series, i.e., the actual warming consists of a linear growth of temperatures imposed on a background of weather noises, as many meteorologists believe now. The discrepancy between the real and artificial temperature dynamics in the series under comparison can be seen even by eye on the interdecadal (8–22 years) scales of both kinds of **WT** patterns. It consists of a considerable weaker variability of the surrogate series over these scales. This subjective estimation is supported by calculations (on the basis of (8) and (9)) of the scaling exponents for both kinds of the series (Table 1). The exponents become slightly positive for the real Northern hemispheric series and rather negative for the Southern Hemisphere. In contrast, all of them are substantially more negative for the surrogate series.

Consequently, we must reject any possibility for the real series to be created as a mixture of a deterministic linear warming trend and pure random noises.

Another, more motivated from the physical point of view, way to create a surrogate climatic series implies consequent adding of the actually observed temperature of 1854 (it is the year of the considered time series start) with shuffled interannual temperature differences of the real series. Such surrogate series model the hemispheric temperature dynamics on the base of the atmospheric heat balance (1) as a kind of **Bm**.

Figures 4 and 5 display two examples of the **Bm**-like series (without any imposed trend and with a linear trend of the 0.6°C per 100 years). The series were created by asynchronous shuffles for both hemispheric interannual temperature differences. Again, one can see tree-like structures of the **WT** patterns. These structures dominate over all scales being considered in the case of the absence of any imposed trend (Fig. 4). However, if a linear trend is imposed, the **WT** patterns can be quite different. For example, the **WT** pattern of the surrogate Northern hemispheric series (see the upper half of Fig. 5) reveals a randomly created intensive interdecadal variation of temperatures. In this case, no signal of the imposed trend can be seen in the largest scales of the **WT** pattern. The example for the Southern Hemisphere does not reveal essential interdecadal temperature variations. In this case, a monotonous growth of temperatures already can be seen (see the bottom half of Fig. 5). Decompositions of both represented **WT** patterns into oscillatory and trend-like components reveal the imposed trend quite clearly.

Calculations give positive values to the scaling exponents in all these examples. Especially large values of the exponents were obtained (see Table 1) for the case of

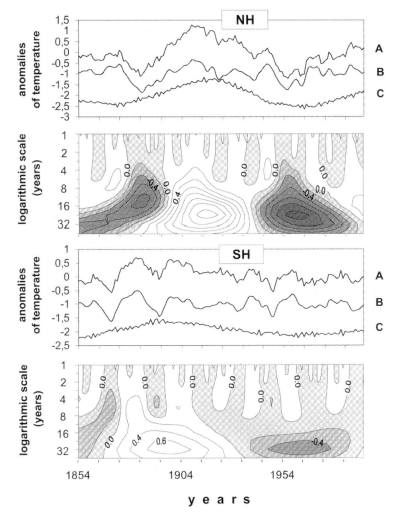

Figure 4
The same as Figure 1 but for the surrogate time series which were formed by random shuffles of the members of the real series that were built on the interannual temperature differences of the primary series from JONES et al. (1994). The starting points of the surrogate series are the actually observed 1854 temperatures.

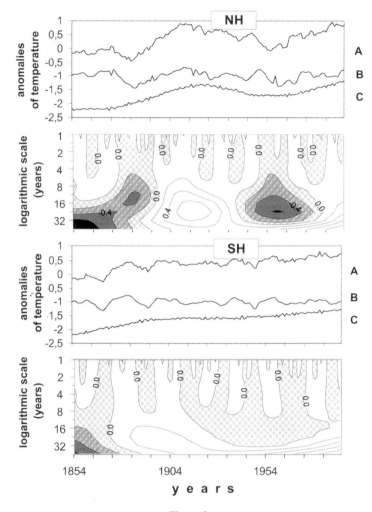

Figure 5
The same as in Figure 4 but for the surrogate series added by the linear trend such as the surrogate series in Figure 3.

Bm in the Northern Hemisphere despite the expected value of the Hurst exponent being 0.5 for all the examples according to the nature of the created ***Bm***. In general, a large dispersion of the scaling exponent estimations seen in Table 1 signals a warning that such estimations for the climatic time series must be considered with great caution. Also it should be noted that, according to the nature of ***Bm***, the multifractal corrections of the scaling exponents must be equal to zero. One can see from Table 1 that most of the calculated corrections deviate from zero quite similarly to such deviations for the real temperature series. Thus, one can conclude that it is impossible to distinguish the mono- and multifractal nature of the real-world climate dynamics on the basis of the hemispheric temperature series under analysis. This pessimistic conclusion is in contrast to published very optimistic findings of multifractality in short and noisy instrumental and even proxy (!) climatic time series (SCHERTZER and LOVEJOY, 1993; SCHMITT et al., 1995).

Comparing the estimations of scaling exponents of the real and surrogate (**Bm**-like) time series one can see that the exponents of the first series are considerably less. Especially for the Southern Hemisphere, these estimations seem to be on an intermediate level between the corresponding estimations for the **Bm** and pure random surrogate series. One reason for that is probably connected with an internally inherent tendency of the real-world climate system to be incessantly equilibrated with respect to external forces, as is depicted by the Langevin form of **Bm** (2). Another reason consists undoubtedly of enormous errors of the hemispheric temperature time series under analysis.

5. Wavelet Transforms of Some Local, Very Extended Air Temperature Time Series

The above considered hemispheric air temperature series are rather short (about 150 data-points only). This circumstance makes it difficult to analyze the character of these series scalings, even over the relatively short time scales, and the more so, the character of the temperature variations of these series behind the crossover scale of 22 years. In order to make such an analysis possible we will consider below several local, but very extended in time (220–338 years) time series based on instrumental records of air temperatures in Europe. Six station records were chosen that form three pairs of the nearest neighbors: Central England (1659–1996) + De-Bilt (1706–1995), Uppsala (1722–1997) + Stockholm (1756–1997), and Vienna (1775–1994) + Prague (1775–1994). This choice is motivated by reason of the evident equality of the scalings of the nearest neighboring stations. Therefore, the differences between the scalings of such stations may be used to measure the statistical tolerance of the scaling exponents estimated by these scalings. The difference between the pairs also may be used to measure the spatial heterogeneity of the scalings.

Figures 6 and 7 show the **WT** patterns of the most extended from the chosen series of Central England and DeBilt series, and the decompositions of these series into oscillatory and trend-like components as well. One can see that the tree-like structures of the displayed **WT** patterns are rather similar. The fact supports the above supposition of the equality of the scalings of the nearest neighbor stations. The scale of 90 years was chosen as the maximal scale in these series decompositions because the tree-like structures extend to the 22–90 scales in their **WT** patterns. Thus, there exist secular temperature variations instead of an apparent monotonous warming trend that was seen in the hemispheric series within these scales. The scalings of these relatively low-frequency variations are quite different from the scalings of the relatively high-frequency temperature oscillations as will be seen below. Among the branches of the tree-like structures it is interesting to consider the domain of negative polarity seen near the left boundary of the Central England **WT** pattern. The domain is very strong. It extends across the large range (from 64 to 1 year) of scales, and corresponds to the famous Little Ice Age that coincides with the Maunder minimum of solar activity. The post-Maunder warming, that is concurrent to the present-day warming, also can be seen in the represented **WT** patterns.

Figures 8–10 show the results of our calculations of the generalized variances (8) for all scales from 1 to 128 years for each pair of stations separately. These calculations confirm the results of our calculations for the hemispheric series: a scaling of the relatively high-frequency temperature oscillations exists over the time scales less than 22 years. Another scaling exists over the longer time scales of 22–90 years. A crossover from the first scaling to the second is seen quite clearly as a sharp bend of each graph inside the scale range of between 16 and 32 years.

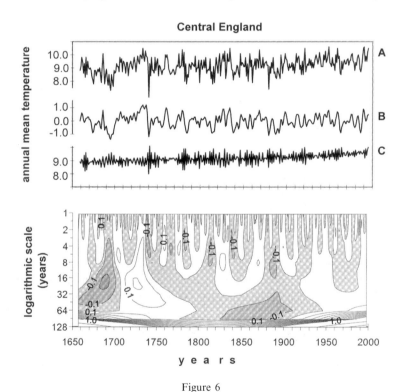

Figure 6
Wavelet transform of the Central England annual mean temperature (1659–1996) series. **A**—the temperature series being transformed, **B**—its stationary oscillatory component evaluated by the inverse wavelet transform (12) with the maximal scale $a_{max} = 90$ years, **C**—the trend-like component evaluated as a residual of subtraction of the oscillatory component from the primary series.

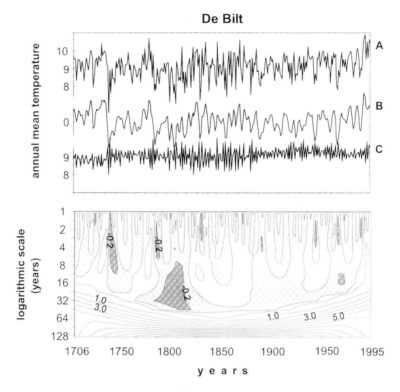

Figure 7
The same as in Figure 6 but for the annual mean DeBilt (1706–1995) temperature time series.

The least-square estimations of the exponent values of both scalings are shown in Tables 2 and 3. Taking into account the tolerances of these values, as they are estimated by the scaling differences within each pair of stations, one can conclude that all exponents of the first scaling are essentially negative (the mean value of the Hurst exponent $H = -0.33$), and those of the second scaling are essentially positive (the mean $H = 1.38$). Thus, there is large contrast in the character of the respective temperature variations.

The relatively high-frequency oscillations look like discontinuous functions of time (see MALLAT, 1998). The first reason for this is quite evident. It is attributable to the numerous and strong noises of the temperature records under analysis. A chaoticity of these oscillations induced by the atmospheric nonlinear dynamics may be presumed as the next reason. However, the first reason seems to be decidedly more essential because such more subtle properties of the nonlinear dynamical oscillations as their multifractality turned out to be suppressed (see Figs. 11–13).

Indeed, there are no trends (signals of the multifractality, see the formula (10)) in the graphs of the ratios $VAR(q, a)/VAR(1, a))^q$ versus a over the range of scales from 1 to 16 years which is the range of the existence of the first scaling. The

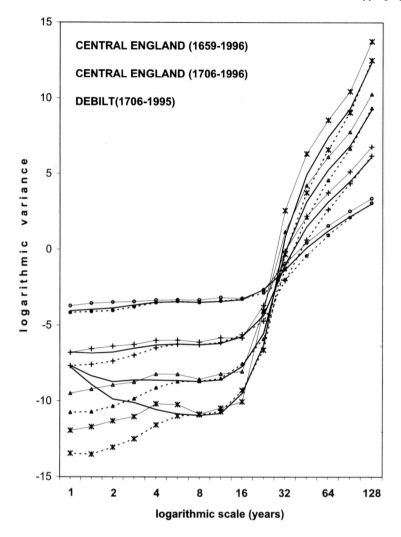

Figure 8
Log-log plots of the generalized variances ($q = 1, 2, 3, 4$) of temperatures versus time scale: the Central England (1659–1996) series (thick solid lines), the same series from 1706 only (thin dotted lines), and the DeBilt (1706–1995) series (thin solid lines).

Central England series (the graphs delineated by thick solid lines in Figures 8 and 11) is a unique exception to the rule. Nonetheless, as we checked, the rejection of the most ancient (1659–1706) part of the Central England series brings the respective graphs (shown by dotted lines) to be quite similar to the graphs of the nearest neighboring DeBilt series (shown by thin solid lines). Thus the unique evidence of the multifractality in the Central England temperature series may be attributed to any specific feature of the temperature oscillations during the Little Ice

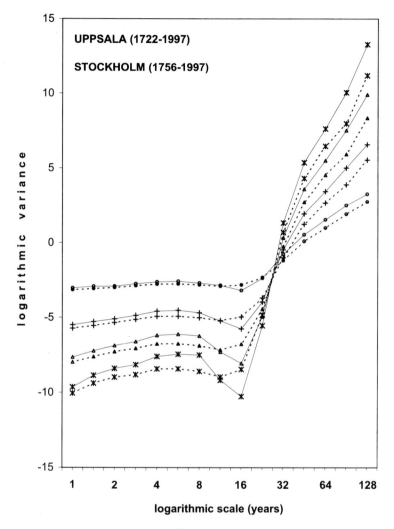

Figure 9
The same as Figure 8 but for the Uppsala (1722–1997) (dotted lines) and the Stockholm (1756–1997) series (solid lines).

Age—Maunder's minimum of solar activity. But still, the recognized multifractality of this climatic event weights against the supposition of a simple, linear response of the real atmosphere to the solar activity forcing, even if such forcing indeed was a trigger of the Little Ice Age.

On the other hand, all relative low-frequency temperature variations have the exponents of their scaling essentially positive and multifractal-like. Their Hurst exponent clearly exceeds the upper limit ($H = 1.0$) that is admissible for the **Bm** processes with long-lived variations of temperature. Also, the popular model of the

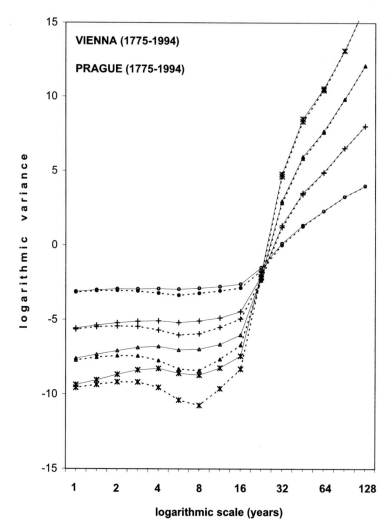

Figure 10
The same as Figure 8 but for the Vienna (1775–1994) series (dotted lines) and the Prague (1775–1994) series (solid lines).

current climate warming which consists of a linear growth of temperature in time as a response to the greenhouse gases concentration growth in the atmosphere, cannot be used to depict well the low-frequency temperature variations under consideration. The Hurst exponent inherent to a monotonous linear warming is equal to $H = 1.0$ although not more. There is nothing for it but to consider the extracted scaling to be a manifestation of a coherent structure of the atmospheric nonlinear dynamics. Such structures have been observed in recent investigations of the fully developed turbulence (see YAMADA and OHKITANI, 1991). In particular,

Table 2

Exponents $\zeta(q)$, $q = 1, 2, 3, 4$ and multifractal corrections $\delta H(q/1)$, $q = 2, 3, 4$ for the first scaling (of high-frequency temperature oscillations) of six local air temperature time series

	$\zeta(1)$	$\zeta(2)$	$\zeta(3)$	$\zeta(4)$	$\delta H(2/1)$	$\delta H(3/1)$	$\delta H(4/1)$
CE (from 1659)	−0.25	−0.54	−1.02	−1.67	−0.04	−0.28	−0.68
CE (from 1706)	−0.33	−0.37	−0.47	−0.56	0.08	0.21	0.36
DeBilt	−0.33	−0.54	−0.69	−0.71	0.13	0.30	0.42
Uppsala	−0.38	−0.72	−1.04	−1.35	0.05	0.11	0.18
Stockholm	−0.46	−0.92	−1.36	−1.81	0.01	0.03	0.04
Vienna	−0.29	−0.54	−0.78	−1.02	0.01	0.11	0.16
Prague	−0.25	−0.47	−0.70	−0.93	0.03	0.05	0.07

a wavelet analysis of turbulence revealed (BENZI and VERGASSOLA, 1991) that such coherent structures can be viewed as events of multifractal processes characterized by very strong (positive or negative) values of the Hurst exponent.

We speculate that the relative low-frequency temperature variations are a mixture of chaotic oscillations and coherent events during the entire time period of the records under analysis. Sometimes (and rather seldom) a coherent jump can be so strong that it is capable of affecting the character of the high-frequency temperature oscillations during the period of this jump. The Little Ice Age seems to be related to such a jump.

6. Conclusions

The detection of oscillatory and trend-like components in data time series in the Earth sciences is an important task in view of both the potential impact of the chaoticity and the nonstationarity in the elaboration of useful models of the processes under analysis.

Table 3

Exponents $\zeta(q)$, $q = 1, 2, 3, 4$ and multifractal corrections $\delta H(q/1)$, $q = 2, 3, 4$ for the second scaling (of low-frequency temperature variations) of six local air temperature time series

	$\zeta(1)$	$\zeta(2)$	$\zeta(3)$	$\zeta(4)$	$\delta H(2/1)$	$\delta H(3/1)$	$\delta H(4/1)$
CE (from 1659)	1.44	2.03	2.46	2.83	−0.85	−1.86	−2.93
CE (from 1706)	1.84	2.67	3.25	3.73	−1.00	−2.25	−3.61
DeBilt	1.45	2.03	2.45	2.82	−0.88	−1.90	−2.98
Uppsala	1.28	1.82	2.16	2.24	−0.73	−1.67	−2.67
Stockholm	1.83	2.00	2.80	3.24	−0.56	−1.28	−2.06
Vienna	1.40	2.09	2.69	3.24	−0.57	−1.29	−2.07
Prague	1.29	2.03	2.61	3.13	−0.55	−1.26	−2.03

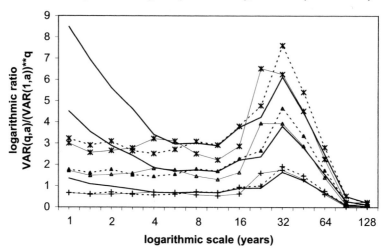

Figure 11
Log-log plots of ratios $VAR(q,a)/(VAR(1,a))^q$ ($q = 2, 3, 4$) versus time scale: The Central England (1659–1996) series (fat solid lines), the same series from 1706 only (thin dotted lines), and the DeBilt (1706–1995) series (thin solid lines).

Here, we considered a relatively new tool of such detection, the wavelet transform of time series, in its application to the well-known complex current climate change problem. It resulted that the wavelet transform actually admits to decompose the time series of interest into statistically stationary oscillations and trend-like components, even if the series are noisy and rather short. The procedure is illustrated on real climatic temperature and surrogate series.

A unique warming trend was extracted from both hemispheric temperature series as a residual of subtractions of inverted **WT**s of these series from the primary series. The more extended local temperature series revealed this trend as part of low-frequency temperature variations over centuries.

To attribute these variations, it proved to be important to estimate the scaling exponents of these temperature series. The estimations do not support the conclusion of some investigations reported earlier concerning the multifractal nature of the climate dynamics in a very wide range of time scales. The relative high-frequency temperature oscillations appear like a monofractal process in our study. It is necessary to rectify the records from heavy observational noises before certain conclusions concerning the multifractality of the high-frequency temperature oscillations become well-grounded.

Heavy observational noises, probably existing in the hemispheric temperature series, also complicate coming to a definite conclusion that the relative low-frequency temperature variations truly represent a **Bm**-process with possible long-lived

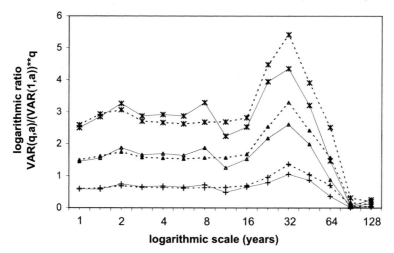

Figure 12
The same as Figure 11 but for the Uppsala (1722–1997) series (dotted lines) and the Stockholm (1756–1997) series (solid lines).

variations of temperatures. Also, the supposition that the current warming consists of a linear (possible externally induced) trend aggravated by weather noises may be rejected for certain. A mixture of chaotic oscillations and coherent structures seems to be more appropriate to model the air temperature dynamics.

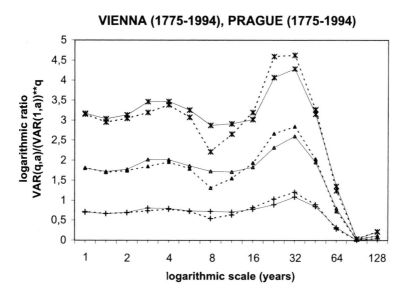

Figure 13
The same as Figure 11 but for the Vienna (1775–1994) series (dotted lines) and The Prague (1775–1994) series (solid lines).

Acknowledgements

Data of the local temperature records used in this study were kindly provided by Drs. D. Parker, K. Schurmans, A. Moeberg, and D. Novotna.

We are particularly thankful to Dr. Yu. P. Koshel'kov for his careful review of the manuscript.

Constructive criticisms of reviewers are also acknowledged.

The initial stage of this research was supported by the Russian Foundation of Basic Research (Grant 94-05-16341).

REFERENCES

BENZI, C., and VERGASSOLA, A. (1991), *Optimal Wavelet Analysis and its Application to Two-dimensional Turbulence*, Fluid Dyn. Res. *8*, 117–126.

BUDYKO, M. I. (1969), *The Effect of Solar Radiation Variations on the Climate of the Earth*, Tellus *21*, 611–619.

DAUBECHIES, I. (1988), *Orthogonal Bases of Compactly Supported Wavelets*, Commun. Pure Appl. Math. *XYI*, 909–996.

ECKMANN, J.-P., and RUELLE, D. (1985), *Ergodic Theory of Chaos and Strange Attractors*, Reviews of Modern Physics *57* (3), Part 1; 617–656.

GROSSMANN, A., KRONLAND-MARTINET, R., and MORLET, J., *Reading and understanding continuous wavelet transform*. In *Wavelets* (Combes, J. M., Grossmann, A., and Tchamitchian, Ph., eds.) (Springer-Verlag, Berlin 1989) pp. 3–20.

HINKLEY, D. V. (1971), *Inference in Two-phase Regression*, J. Am. Stat. Assoc. *66*, 736–743.

HOUGHTON, J. T., MEIRA FILHO, L. G., CALLANDER, B. A., HARRIS, N., KATTENBERG, A., and MASKELL, K. (eds.), *Climate Change 1995. The Science of Climate Change*, IPCC 1996 (Cambridge Univ. Press, Cambridge 1996) 572 pp.

JONES, P. D., WIGLEY, T. M. L., and BRIFFA, K. R., *Global and hemispheric temperature anomalies—Land and marine instrumental records*. In *TRENDS'93: A Compendium of Data on Global Change* (Boden, T. A., Kaiser, D. P., Sepanski, R. J., and Stoss, F. W. eds.) (ORNL/CDIAC-65, Oak Ridge Nat. Lab., Tenn., USA 1994) pp. 603–608.

MALLAT, S., A *Wavelet Tour of Signal Processing* (Boston, Academic Press 1998) 700 pp.

MANDELBROT, B. B., *The Fractal Geometry of Nature* (San Francisco, Freeman 1982) 465 pp.

MEYER, Y., *Orthogonal wavelets*. In *Wavelets* (Combes, J. M., Grossmann, A., and Tchamitchian, Ph., eds.) (Springer-Verlag, Berlin 1989) pp. 21–37.

MEYERS, S. D., KELLY, B. G., and O'BRIEN, J. J. (1993), *An Introduction to Wavelet Analysis in Oceanography and Meteorology: with Application to the Dispersion of Yanai Waves*, Mon. Wea. Rev. *121*, 2858–2866.

RIOUL, O., and FLANDRIN, P. (1992), *Time-scale Energy Distributions: A General Class Extending Wavelet Transforms*, IEEE Transactions Signal Processing *40*, 1746–1757.

SCHERTZER, D., and LOVEJOY, S. (1993), *Nonlinear Variability in Geophysics 3. Scaling and Multifractal Processes*, Lecture Notes (Institut d'Études Scientifiques de Cargèse, 10–17 Sept. 1993) 292 pp.

SCHMITT, F., LOVEJOY, S., and SCHERTZER, D. (1995), *Multifractal Analysis of the Greenland Ice-core Project Climate Data*, Geophys. Res. Lett. *22*, 1689–1692.

SCHLESINGER, M. E., and RAMANKUTTY, N. (1994), *An Oscillation in the Global Climate System of Period 65–70 Year*, Nature *367*, 723–726.

SOLOV, A. R. (1988), *A Bayesian Approach to Statistical Inference about Climate Change*, J. Clim. *1*, 512–521.

SONECHKIN, D. M., DATSENKO, N. M., and IVACHTCHENKO, N. N. (1996), *New method of chaotic time series extrapolation by means of wavelets with an application to the climate dynamics*, Isvestija Vistchich Utchebnich Savedeny, Pricladnaja Nelineinaja Dinamika (Applied Nonlinear Dynamics) *4*, 108–121, in Russian.

SONECHKIN, D. M., DATSENKO, N. M., and IVACHTCHENKO, N. N. (1997), *Estimation of the Global Warming Trend by Wavelet Analysis*, Izvestiya AN, Fizika Atmosfery i Okeana *33*, 184–194. English Translation: Izvestiya, Atmospheric and Oceanic Physics *33*, 167–176.

WOODWARD, W. A., and GRAY, H. L. (1993), *Global Warming and the Problem of Testing for Trend in Time Series Data*, J. Clim. *6*, 953–962.

YAMADA, M., and OHKITANI, K. (1991), *Orthogonal Wavelet Analysis of Turbulence*, Fluid Dynamics Research *8*, 101–115.

(Received March 5, 1998, revised June 7, 1998, accepted June 23, 1999)

To access this journal online:
http://www.birkhauser.ch